PREPARING FOR THE NATE EXAM

Gas and Oil Heating

ILLUSTRATION CREDITS
FRONT COVER, LEFT TO RIGHT: ISTOCKPHOTO, ISTOCKPHOTO, VEER, ISTOCKPHOTO
OVERLEAF: VEER
ALL PHOTOS USED WITH PERMISSION.

ISBN-13: 978-1-61607-164-6
ISBN-10: 1-61607-164-8

Third printing.

Introduction

PREFACE

What is the purpose of this book?

This book has two purposes—to serve as a valuable research and reference source for technicians in the HVACR industry, and to offer solid, in-depth assistance for those technicians preparing to take the NATE Certification Exam. This publication directly reflects the continuing efforts of major industry organizations (including ACCA, NATE, PHCC, and RSES) to promote and encourage a single certification program for technicians in the industry. As such, it assumes a basic working knowledge of HVACR systems on the part of the reader. It is not intended to be exhaustive, but individuals who want to take a specific NATE Exam can focus their study on the subject matter contained in this book (or in other books in this series) to help them prepare for the test. For more comprehensive information, please visit www.rses.org, where you will find many other resources, including RSES training courses that expand greatly on the material included here.

What is NATE?

As an independent, third-party, nonprofit certification body supported by a broad-based coalition of industry leaders, North American Technician Excellence, Inc. (NATE) represents a shared commitment to improving the HVACR industry through voluntary testing and certification. NATE supporters include contractors, distributors, education and training providers, manufacturers, technicians, utilities, and their respective trade associations. There are 21 NATE installation, service, and senior tests. Specialties include Air Conditioning, Air Distribution, Heat Pumps, Gas Furnaces, Oil Furnaces, Hydronics Gas, Hydronics Oil, Light Commercial Refrigeration, Commercial Refrigeration, and Senior HVAC Efficiency Analyst.

Why is it so important to be NATE-certified?

Your NATE certification tells the world that you have the know-how and ability to do the job right. It marks your status as a professional whose high work standard is officially recognized and documented. Because it signifies that you are highly competent and knowledgeable in your field, your employer and customers will quite likely consider your certified status as being an indication that your services are more valuable. They know that your ability to perform high-quality service will benefit them, and they'll be willing to compensate you accordingly. Your standing as a certified technician also lends added prestige to the entire industry—and that comes back as a benefit for you, through increased recognition and respect.

Any tips on how I can prepare for the NATE Exam?

Begin well ahead of time to prepare for the test by using this book as a study aid. Find a quiet place where you can study without distraction. Learning is a process that requires reflection, so try to avoid waiting until the last minute to study. Retention of reading material increases greatly when you spend at least as much time *thinking* about the material as you do actually *reading* the text. Research relating to test scores indicates that studying in an environment that duplicates the testing environment as closely as possible may improve your recall of information during the test. If possible, try to sit in a chair at a desk or table during study periods. Make sure that your study area has good lighting. Get enough sleep the night before the Exam, so you'll be fresh and alert on the day of the test.

Other helpful information you should know:

- You may bring a non-programmable calculator with you to use during the test, but *no cell phones or other electronic devices are permitted* in the test area. Scratch paper and pencils are provided.
- You may take a NATE Certification Exam in either pencil-and-paper or electronic (computer) form. Regardless of the format you choose, *all NATE testing is proctored.*
- Make sure that you have photo identification with you. Arrive well ahead of the test start time to give yourself a chance to get comfortable, so that you can feel relaxed and able to focus on the Exam.

At the end of each Chapter in this book are some Review Questions to help you test your knowledge. The answers to these questions appear in an Appendix at the back of the book for your convenience. Good study habits increase your chance of earning a good score on the Exam. However, it is important to remember the key reason for studying the material in this book. Your main goal should be to master the concepts being studied so that you can improve your skills and earn your standing in the trade community as a master technician or installer of HVACR systems.

ACKNOWLEDGEMENTS

RSES owes a debt of gratitude to the many people who assisted in the development and preparation of this study material. The untiring efforts of these dedicated individuals are sincerely appreciated. In particular, RSES wishes to thank:

- Roger Hensley, CMS, Educational and Examining Board Chairman
- Ratib Baker, CMS
- Rolf Blom, CMS
- Hugh Cole, CMS
- Rich Hoke, CMS
- Scott Nelmark
- Nick Reggi, CMS
- Bill Sammons, CMS
- Loren Shuck, CMS

Table of Contents

Chapter 1
Introduction to Gas and Oil Heating. 1
Chapter 2
Combustion 9
Chapter 3
Furnace Installation 23
Chapter 4
Venting. 37
Chapter 5
Air Flow 53
Chapter 6
Troubleshooting 65
Chapter 7
Oil Burners 87
Chapter 8
Oil Tanks 103
Chapter 9
Planned Maintenance. 111
Appendix A
Answers to Review Questions. 125

CHAPTER 1

Introduction to Gas and Oil Heating

INTRODUCTION

Most of today's residential and light commercial fossil fuel heating systems use either gas or oil as the fuel source. Much of the information in this book is relative to both types of systems. There are some differences between gas and oil, of course, and those differences are noted so that the technician who wishes to become NATE-certified can prepare for either the Gas Heating exam or the Oil Heating exam. The final three Chapters in this book are devoted to oil systems.

A modern furnace uses a heat exchanger to transfer heat from the combustion chamber to the supply air that heats the conditioned space. Some early furnaces, occasionally called "gravity" furnaces, moved the supply air by means of *natural convection* (based on the principle that warm air rises), but today's typical forced-air furnace uses a fan to move the supply air across the heat exchanger. Most heat exchangers are made of steel, often coated with another material to reduce corrosion. This allows heat to transfer quickly, while preventing flue gases from getting into the airstream and entering the living space.

Higher-efficiency furnaces with AFUE (*annual fuel utilization efficiency*) ratings in the 90%+ range add a secondary heat exchanger (see Figure 1-1). The secondary heat exchanger draws additional sensible heat (reducing the temperature) and latent heat (condensing moisture from the flue gases). Since the flue gases are cooled sufficiently to condense out a significant amount of water, the secondary heat exchanger needs to be able to withstand the corrosive effects of the acidic water from the flue gases. To prevent corrosion, secondary heat exchangers may be made of stainless steel, coated steel, or even high-temperature plastics.

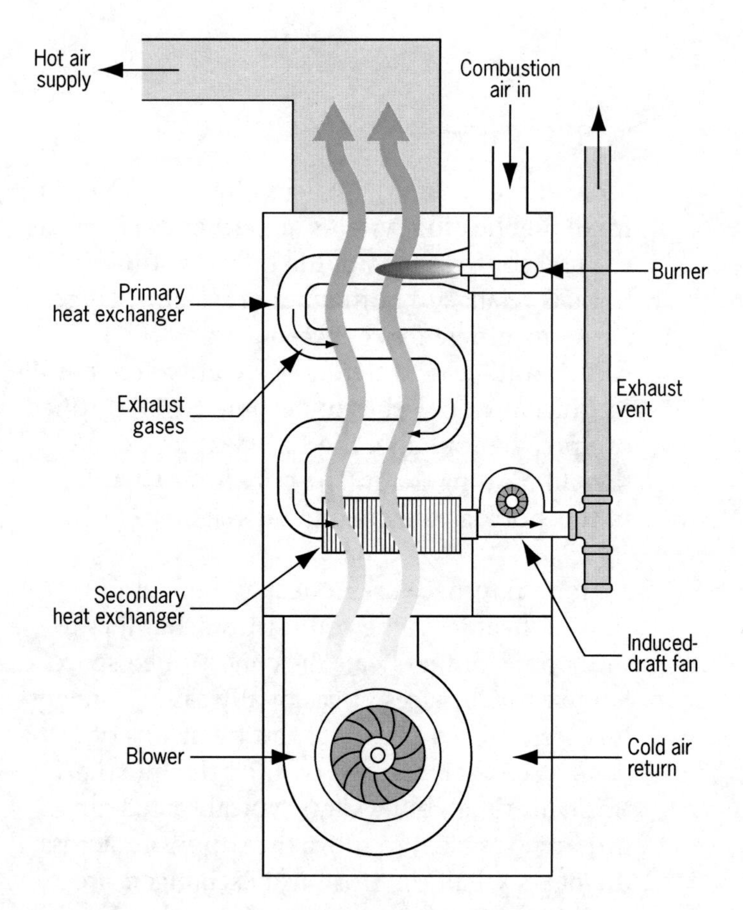

Figure 1-1. High-efficiency furnace with two heat exchangers

FUELS

Gas

There are three types of gases that are commonly used for heating purposes:

- natural gas
- liquefied petroleum (or LP) gas
- manufactured gas.

To some extent, all of these gases are blends. Gas suppliers blend different types of gases in an effort to get consistent performance and Btu output from the final product. *Natural gas*, for example, is a blend consisting mostly of methane, with small amounts of propane added to balance the Btu content of the gas. It is usually provided by a utility company and delivered via a local pipeline to the building. *LP gas* is a blend made primarily of propane, with butane added to balance the Btu content. It is used in areas where a connection to a pipeline is not available. *Manufactured gas* is used to a lesser degree. It is a blend of butane and other gases that are often produced as by-products of various manufacturing processes. Both LP gas and manufactured gas must be delivered by truck and stored in a tank on the property.

It is important to note that while natural gas and manufactured gas are lighter than air and will dissipate quickly with good ventilation, LP gas is heavier than air and will settle out in low spots. This is an added safety concern when working with LP gas. Types of gases and their characteristics are further described in Table 1-1.

	Natural gas	Manufactured and mixed gases	Liquid petroleum (LP) gas
Contents	90% methane	Variable hydrocarbons	Primarily propane
Specific gravity	0.60 (lighter than air)	0.60 (lighter than air)	1.52 (heavier than air)
Heat content	≈1,050 Btu/ft^3	≈500 to 800 Btu/ft^3	≈2,500 Btu/ft^3
Manifold pressure	3.5 in. w.c.	2.5 in. w.c.	Up to 11 in. w.c.
Combustion air per cubic foot of gas	10 ft^3	10 ft^3 or less	24 ft^3

Table 1-1. Types of gas fuels

Oil

Fuel oil is available in different grades, numbered from 1 to 6 according to boiling point, flash point, viscosity, and other characteristics. For residential and light commercial heating purposes, No. 2 fuel oil is the most widely used. This grade of oil is a *distillate*, which is very fluid at room temperature. It has a heating value of 137,000 to 142,000 Btu/gal. Heating oil used in domestic and small commercial applications is commonly delivered by truck to on-site storage tanks. The oil is drawn from the tank by a pump and *atomized* into a fine spray by forcing it under pressure through a nozzle. The spray is then ignited by an electric spark. *Biofuels*, which are blends of conventional heating oil and biodiesel, also are used in some heating applications. Biofuels are labeled as B5, B10, B15, etc. The number indicates the percentage of biodiesel present in the blend (i.e., B5 contains 5% biodiesel, B10 contains 10% biodiesel, and so on).

GAS PRESSURES

Pressure is the force per unit area exerted by a gas or liquid on the walls of its container. In

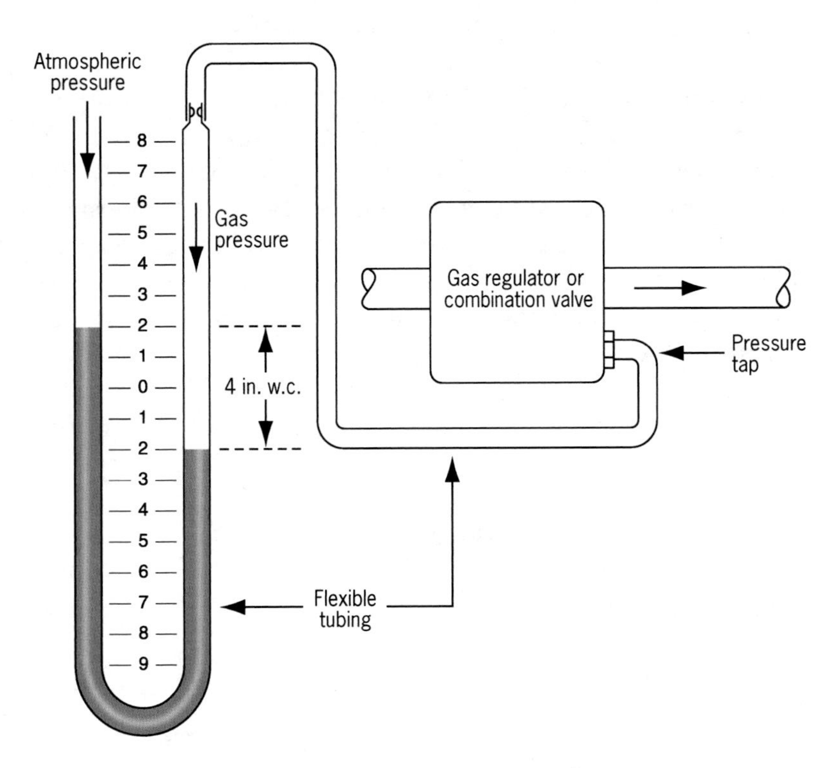

Figure 1-2. U-tube manometer measuring pressures

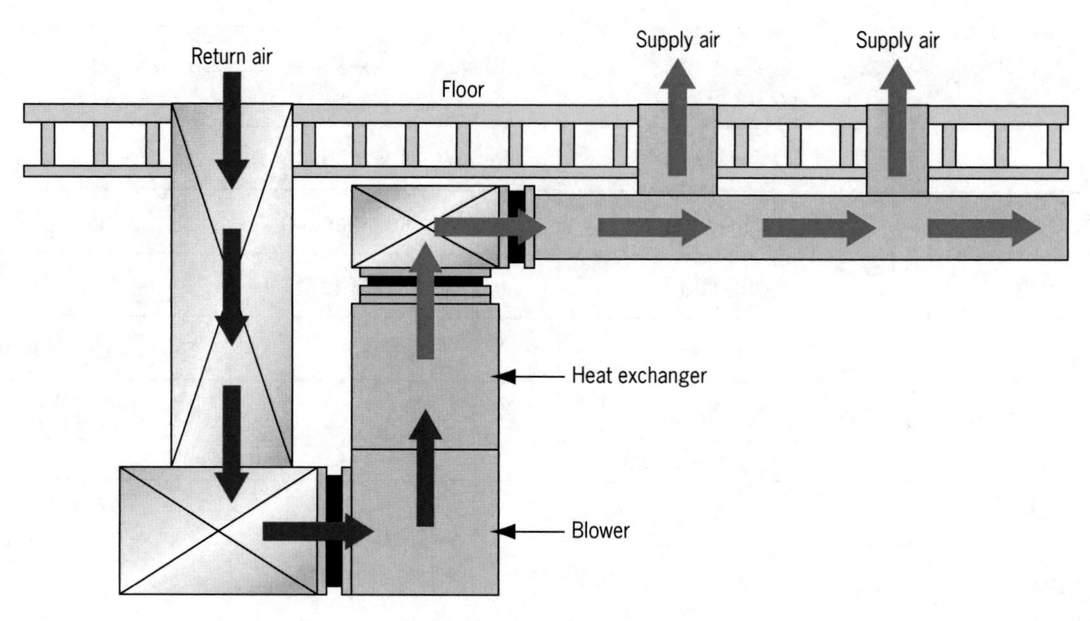

Figure 1-3. Upflow furnace

gas appliance servicing, pressures are usually measured (with a manometer) in inches of water column (in. w.c.). Figure 1-2 on the previous page shows how a U-tube manometer is used to measure pressures in a gas heating system.

Natural gas normally is delivered at a pressure of no more than 0.5 psi, or approximately 14 in. w.c. A higher pressure can damage the regulators in the equipment gas valves. The regulators at the unit reduce the gas pressure to the proper operating pressure. (Some larger commercial buildings may require pressures of 2 psi or even 5 psi, due to the volume of gas being used and the length of pipe being run. In such situations, a second regulator is needed near each unit to step the pressure down to less than 14 in. w.c.) The typical manifold pressure for natural gas is 3.5 in. w.c., but equipment varies and it is important always to check the manufacturer's specifications. Some models even run a negative gas pressure.

LP gas appliances commonly operate at higher pressures than comparable natural gas appliances. Due to the higher Btu/ft^3 content of LP gas, orifices are usually smaller for the same Btuh input. LP gas often operates at higher manifold pressures as well (up to 11 in. w.c.), but the proper operating pressure should be verified with the manufacturer's requirements, since some LP gas appliances equipped with power burners are designed to operate at the same 3.5 in. w.c. as natural gas furnaces. The higher orifice pressures generally used with propane increase primary air injection and assist in moving the heavier-than-air mixture of gas and air through the burner ports. Natural gas furnaces that are to be used with LP gas may need to have the orifices (both main and pilot) changed and the regulator in the gas valve replaced or modified. In all modifications, it is important to follow the manufacturer's instructions and use only parts supplied or approved by the manufacturer.

FURNACE CONFIGURATIONS

Furnaces are available in many different configurations. The type of structure in which the unit is installed typically dictates the style of furnace used. Some common configurations and their applications are discussed below.

Upflow furnaces normally are used in homes that have basements, or in closet installations with overhead duct runs and ceiling diffusers. In an upflow furnace, the cool return air enters the furnace at the bottom and the heated supply air exits through the top of the furnace for distribution into the home (see Figure 1-3). Depending on space requirements, an upflow furnace frequently offers the easiest addition of an evaporator coil for central air conditioning.

As you might expect, the air flow pattern in a *downflow* (or *counterflow*) furnace is the opposite of an upflow furnace. That is, the return air is drawn in through the top of the furnace and the conditioned air is discharged from the bottom of the furnace (see Figure 1-4 below). Downflow furnaces often are installed in closets or utility rooms, with the duct system embedded in a poured concrete slab or suspended under the floor in a crawl space.

Horizontal furnaces frequently are installed in attics and crawl spaces, where vertical space (height) is limited. Return air enters one end of the furnace, and conditioned supply air is expelled from the opposite end (see Figure 1-5 on the next page). Horizontal furnaces typically are configured by the manufacturer for air to flow in one direction—from left to right, for example—although some manufacturers allow field conversion to reverse air flow.

Lowboy furnaces were designed to have a lower profile than standard furnaces, and are typically installed in crawl spaces or in basements with low ceilings. They are usually less than 4 ft tall. Both the return and supply ducts are connected to the top of the unit (see Figure 1-6 on the next page). The blower is behind the furnace in a separate

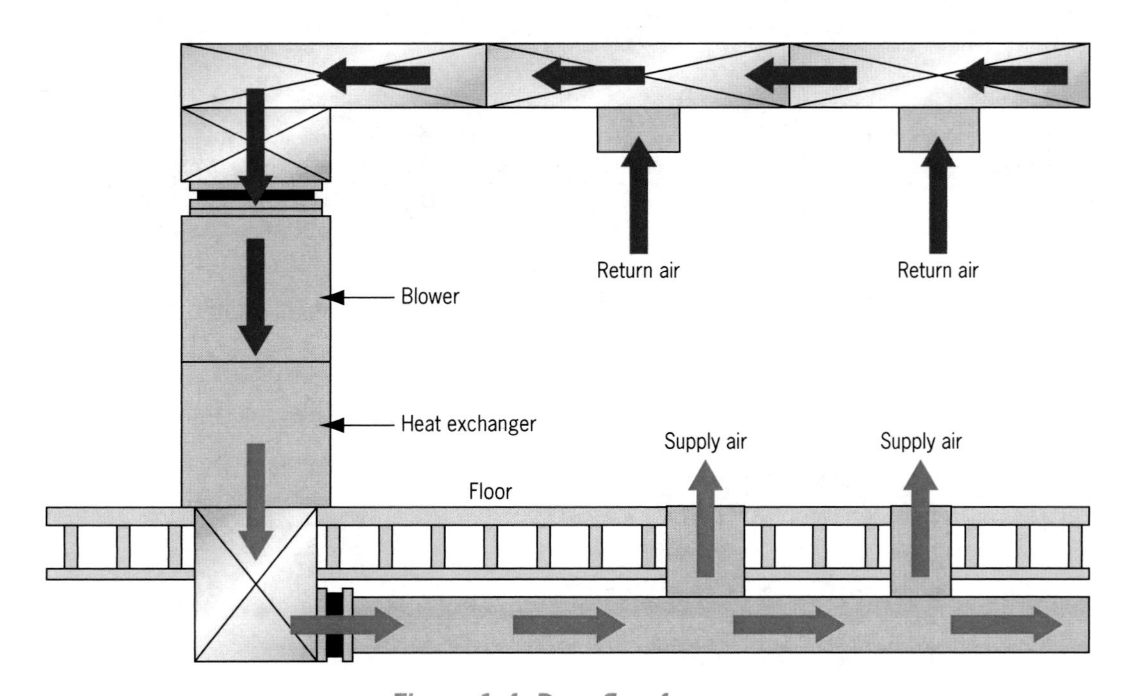

Figure 1-4. Downflow furnace

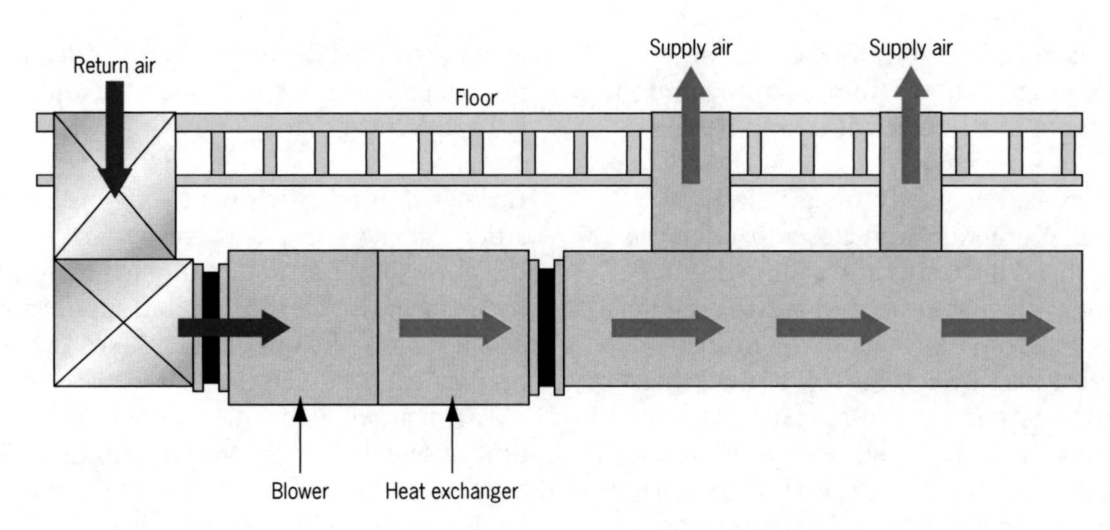

Figure 1-5. Horizontal furnace installed in crawl space

compartment. Today's horizontal furnaces and the smaller size of equipment in general have for the most part eliminated the need for the lowboy design.

A *multipositional* furnace may be installed in the upflow, downflow, or horizontal position. These units generally are shipped from the factory in the upflow configuration. When changing the furnace configuration, the installer may be required to make changes internal to the furnace. Be sure to follow the furnace manufacturer's recommendations exactly regarding application in various positions.

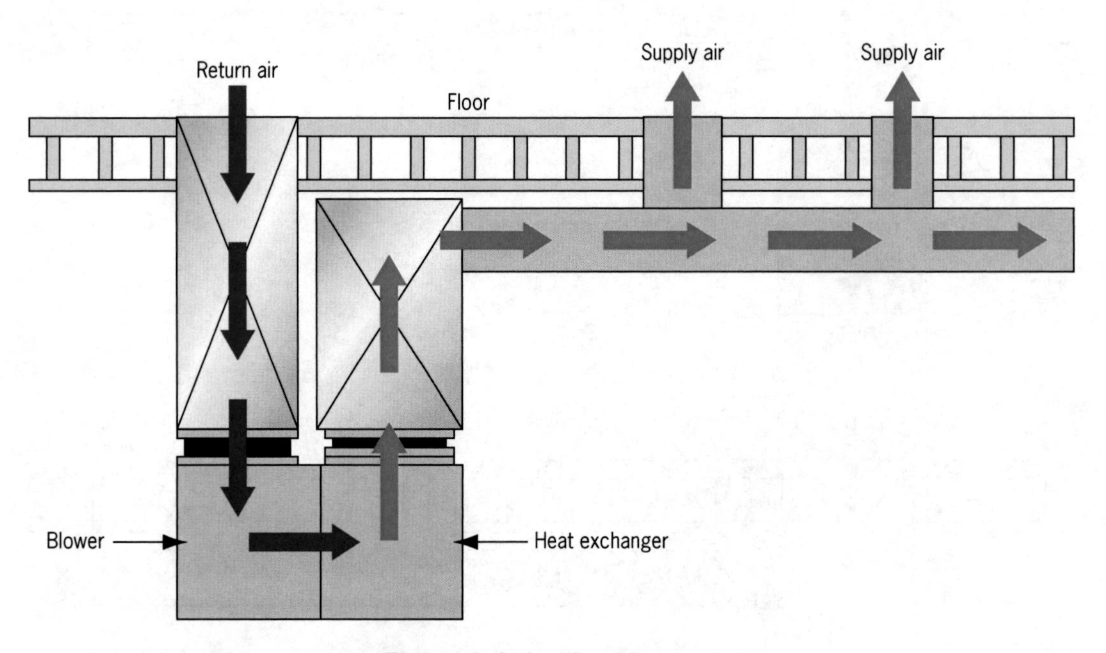

Figure 1-6. Lowboy furnace

In all configurations, the products of combustion (burned fuel) are separated from the air circulated to the home by the heat exchanger.

The key thing for you as a service technician to remember is that the sequence of operation remains essentially the same for all furnaces, even though the configurations may differ. Once you are familiar with the components, their proper set-up, and how they function within the heating cycle, you will be able to service all types of furnaces effectively. □

REVIEW QUESTIONS

1. In a modern gas furnace, the component that separates the combustion gases from the supply air is the __________.

   a. gas valve
   b. heat exchanger
   c. inducer fan
   d. plenum box

2. What type of furnace moves warm air through the conditioned space without a fan?

   a. Gravity
   b. Horizontal
   c. Lowboy
   d. Upflow

3. Furnace heat exchangers generally are made of __________.

   a. ceramic
   b. copper
   c. plastic
   d. steel

4. What does AFUE stand for?

   a. Actual Fuel Usage Efficiency
   b. Annual Fuel Utilization Efficiency
   c. Annual Furnace Usage Equivalency
   d. Approved Fuel Usage Estimate

5. The main purpose of a secondary heat exchanger is to __________.

   a. evaporate the flue gases
   b. provide a backup containment
   c. remove additional heat from the flue gases
   d. reverse the primary air flow

6. Which gas commonly used as a fuel in gas furnaces is heavier than air?

   a. Butane
   b. Isobutene
   c. Methane
   d. Propane

7. Which grade of fuel oil is most commonly used in residential furnaces?

   a. 1
   b. 2
   c. 3
   d. 4

8. How many cubic feet of air are needed to burn 1 ft^3 of natural gas?

   a. 6 ft^3
   b. 10 ft^3
   c. 18 ft^3
   d. 24 ft^3

9. How many cubic feet of air are needed to burn 1 ft^3 of propane?

   a. 6 ft^3
   b. 10 ft^3
   c. 18 ft^3
   d. 24 ft^3

10. What is the approximate heat content of 1 ft^3 of natural gas?

   a. 500 Btu
   b. 800 Btu
   c. 1,050 Btu
   d. 2,500 Btu

11. What is the approximate heat content of 1 ft^3 of propane?

   a. 500 Btu
   b. 800 Btu
   c. 1,050 Btu
   d. 2,500 Btu

12. What is the approximate heat content of one gallon of No. 2 fuel oil?

   a. 75,000 to 90,000 Btu
   b. 95,000 to 135,000 Btu
   c. 137,000 to 142,000 Btu
   d. 145,000 to 170,000 Btu

13. What is a typical manifold pressure for a natural gas furnace?

   a. 3.5 in. w.c.
   b. 11 in. w.c.
   c. 2 psi
   d. 5 psi

14. What is a typical atmospheric manifold pressure for a propane furnace?

   a. 3.5 in. w.c.
   b. 11 in. w.c.
   c. 2 psi
   d. 5 psi

15. What type of furnace is commonly installed in a closet with underfloor supply ducts?

   a. Downflow
   b. Horizontal
   c. Lowboy
   d. Upflow

CHAPTER 2

Combustion

COMBUSTION CHEMISTRY

Combustion is a chemical reaction. When a source of ignition is present, the process of combining fuel and oxygen results in the release of heat. Fuels are made up of varying amounts of hydrogen and carbon. When hydrogen (H) and carbon (C) are combined with the oxygen (O) in the air, the by-products are carbon dioxide (CO_2), water vapor (H_2O), and heat. Three things must be present in order for combustion to take place: *fuel*, *oxygen*, and *heat*. These three ingredients frequently are referred to as the "fire triangle" (see Figure 2-1). Remove any one of these three components and the fire will go out.

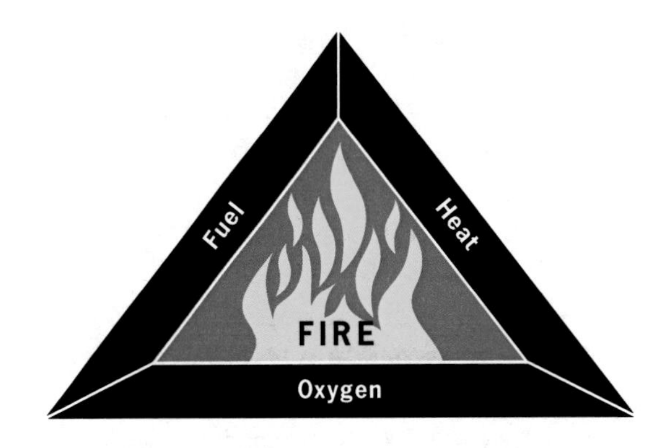

Figure 2-1. Fire triangle: fuel + heat + oxygen = fire

Once the mixture of fuel and oxygen is ignited, the chemical reaction of combustion supplies sufficient heat of its own for the process to continue. The chemical change that takes place in the *complete* combustion process can be expressed in the form of a balanced chemical equation. For methane, the equation is written as follows:

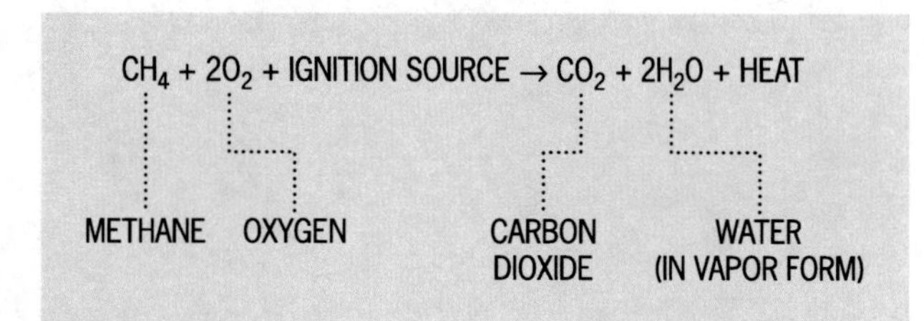

If an insufficient amount of oxygen is mixed with the fuel or if the temperature is cooled by flame impingement on the cooler surfaces of the heat exchanger, *incomplete* combustion can occur. Incomplete combustion results in the formation of carbon monoxide (CO), soot, and other by-products. The two sides of the chemical equation cannot be balanced. For example:

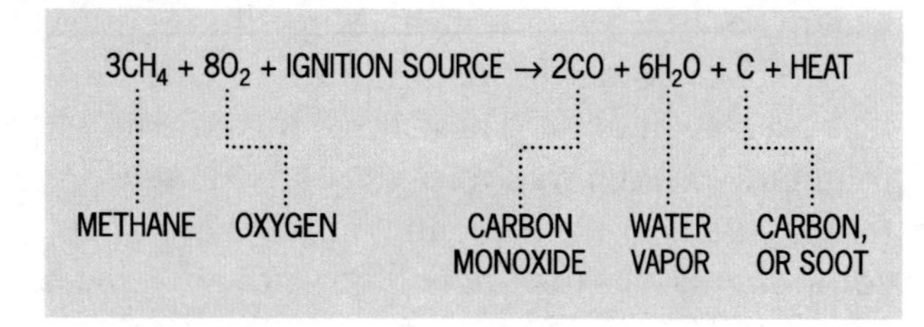

QUANTITY AND QUALITY OF COMBUSTION AIR

Regardless of the location of a furnace, adequate combustion air must be available at all times. In older homes, the construction was much looser and the furnace often was installed in what would be considered an *unconfined space* (such as a basement). In accordance with National Fire Protection Association (NFPA) Standard 31 (*Standard for the Installation of Oil-Burning Equipment*), NFPA 54 (*National Fuel Gas Code*), and NFPA 58 (*Liquefied Petroleum Gas Code*), an "unconfined space" is defined as any space whose volume is equal to or greater than 50 ft^3 per 1,000 Btu of input rating (or 20 Btu/ft^3). In an unconfined space, a furnace draws its combustion air from the space directly around the furnace.

In today's homes, construction is tighter and more furnaces are being placed in smaller spaces. As a result, it is more important than ever to ensure that proper combustion air is supplied to the furnace. In many applications, this means that additional combustion air must be brought into the furnace area. If sufficient combustion air is not available, incomplete combustion will result.

The amount of combustion air required for a given space is calculated based on the total input rating of all fuel-burning appliances installed in that space. Only areas that "freely communicate" with the space (e.g., areas that have fully louvered doors or no doors) can be considered part of an unconfined space. If the actual free area of a louvered door is not known, wood louvers are assumed to have a 20 to 25% free opening. Metal louvers or grilles are assumed to have a 60 to 70% free opening.

Look at the floor plan of the boiler room shown in Figure 2-2. To determine the maximum total input firing rate allowable in this room, first calculate the volume as follows:

$$\begin{aligned} &(30\text{ ft} \times 20\text{ ft} \times 8\text{ ft}) = 4{,}800\text{ ft}^3 \\ \text{minus } &(10\text{ ft} \times 10\text{ ft} \times 8\text{ ft}) = \underline{800\text{ ft}^3} \\ &= 4{,}000\text{ ft}^3 \end{aligned}$$

Then:

$$\frac{4{,}000\text{ ft}^3 \times 1{,}000\text{ Btu}}{50\text{ ft}^3} = 80{,}000\text{ Btu}$$

$$\frac{80{,}000\text{ Btu} \times 1\text{ gal/hr of No. 2 fuel}}{140{,}000\text{ Btu}} = 0.57\text{ gal/hr}$$

Result: If you fire greater than 0.57 gal/hr or 80,000 Btu, you need additional combustion air.

If air from an adjacent room is being added to a confined space (such as a utility closet) to provide combustion air, two openings between the rooms must be made, one 12 in. above the floor and one 12 in. below the ceiling. The size of the openings is based on an input of 1 in^2 per 1,000 Btu.

If air is added directly from the outside of the structure, two openings, one near the floor and one near the ceiling, must be made. The size of these openings is based on an input of 1 in^2 per 4,000 Btu. The above requirements are set forth in NFPA 31 and NFPA 54/58 (current editions). Alternatively, if operating in a confined space, additional air may be added by a duct to the outside, sized on an input of 1 in^2 per 5,000 Btu. This duct should terminate with two openings, one near the floor and one near the ceiling.

Combustion air that is contaminated with chemicals, dust, bleach, dyes, paint fumes, hair spray, dryer lint, etc. can foul pilot assemblies and burners and result in detrimental flue products that will lead to corrosion and failure of the heat exchangers. Areas with significant amounts of contamination benefit from direct vent systems that draw combustion air from the outside.

Direct vent systems draw combustion air directly from the outside to a sealed combustion chamber and then vent back to the outside. Both the intake and the exhaust for this type of system must draw from the same area so that they are at equal pressures. This type of system is commonly used with modern 90% furnaces. All of the manufacturer's instructions relating to the installation of direct vent systems *must* be followed. More information on venting is provided in Chapter 4.

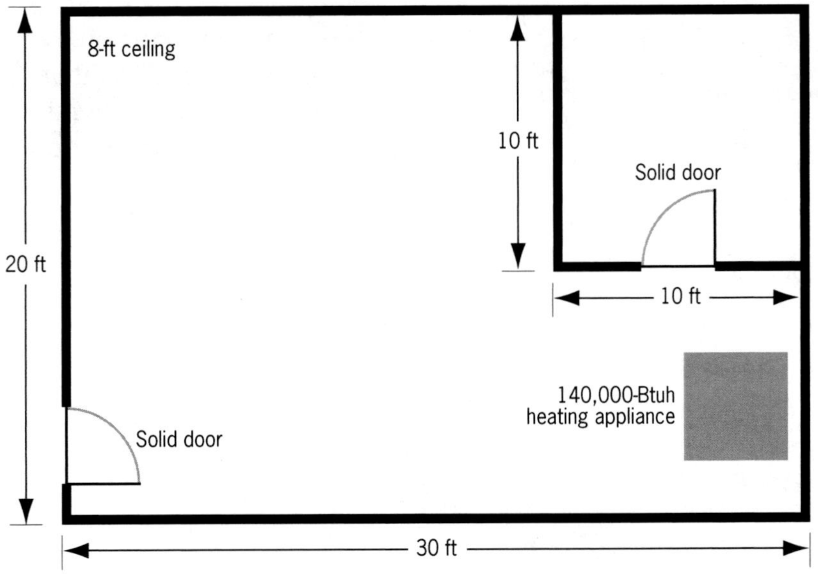

Figure 2-2. Calculating combustion air

GAS BURNERS

Gas furnaces use several different types of burners. Two of the most common are ribbon burners and inshot burners (see Figure 2-3 on the next page). *Ribbon* burners are generally long and narrow so that they can be positioned from front to back inside the bottom of the heat exchanger. These types of burners often are used in atmospheric systems with natural-draft flue vents.

When ribbon burners are removed for cleaning, they must be reinstalled in the correct order and alignment to ensure that the flames do not impinge against the heat exchanger or other burners. The crossover bars must carry the flame from burner

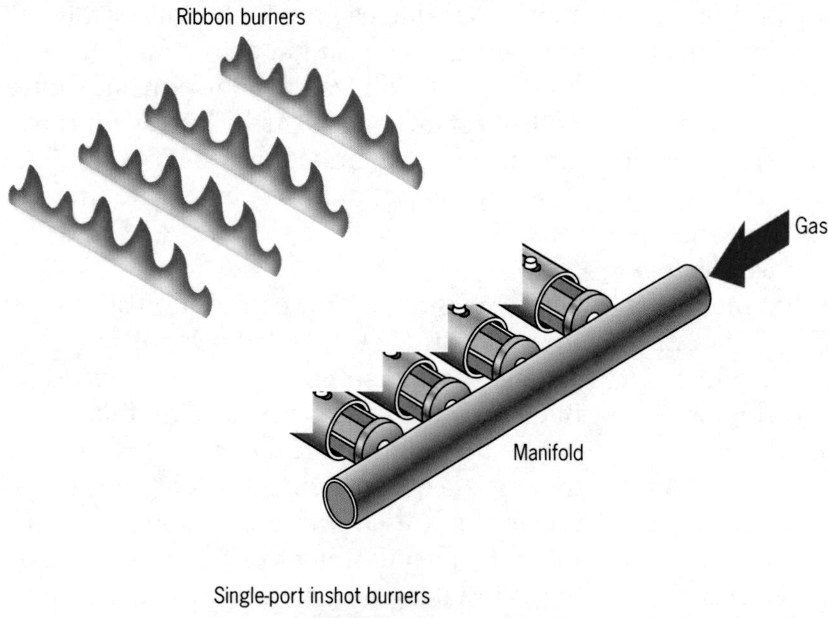

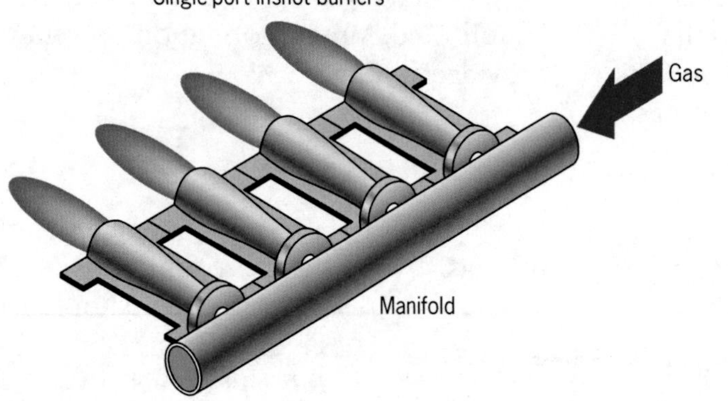

Figure 2-3. Common gas burner types

to burner for a smooth light-off. Ribbon burners provide relatively quiet operation.

Inshot burners frequently are used on newer 80% and 90% furnaces in combination with induced-draft fans that draw the flue gases through the heat exchanger. Inshot burners are a little noisier than ribbon burners, with a relatively short, hard flame that burns just inside the inlet of the heat exchanger. For both types of burners, the flames should be a light blue color, with little or no yellow at the tips.

Figure 2-4 shows a typical atmospheric gas burner and its major parts. An *orifice* in the gas manifold directs and regulates the flow of gas into the burner, as shown in Figure 2-5. The velocity of the gas flow into the burner creates a venturi effect that draws in primary air to mix with the gas before ignition.

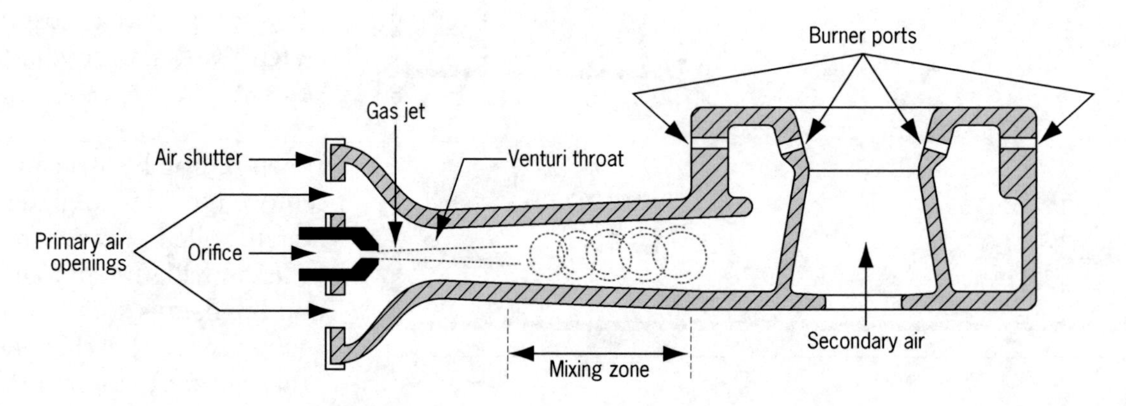

Figure 2-4. Typical atmospheric gas burner

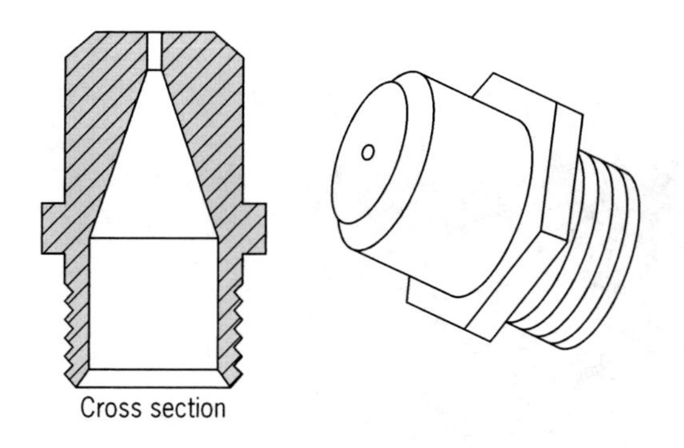

Figure 2-5. Orifice through which gas enters burner

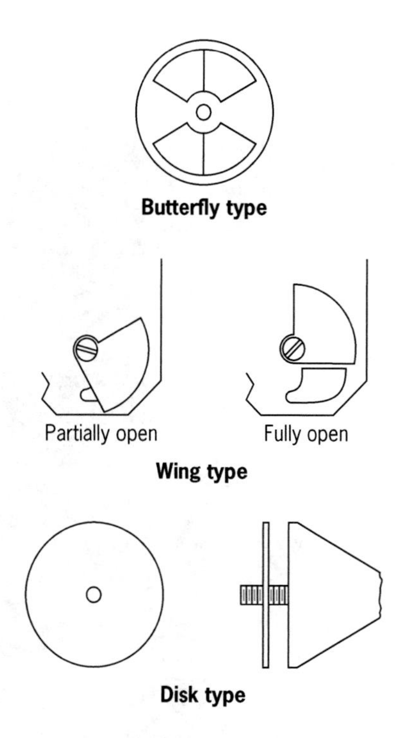

Figure 2-6. Air shutters

Most atmospheric burners have some type of *air shutters* (see Figure 2-6) at the inlet of the burner to control the flow of primary air. Primary air in an inshot burner is controlled by the gas pressures and the induced-draft fan. The inshot burner does not have a primary air adjustment. Secondary air is added to both types of burners after ignition to complete the combustion process.

GAS BURNER COMBUSTION EFFICIENCY TESTING

Before attempting to check gas burner combustion efficiency, make sure that you have the proper testing tools. Figure 2-7 on the next page shows an older test kit, which consists of a manometer (15 in. w.c.), a CO_2 indicator with aspirator, a draft gauge (–0.10 to +0.14 in. w.c.), a stack thermometer (0 to 1,000 °F, 5½-in. stem), and a fire efficiency slide rule. The "dumbbell" style CO_2 tester requires no batteries to operate. It has been used for many years and has proven to be an accurate and reliable tester. It contains a fluid (potassium hydroxide) that absorbs the CO_2 in the gas sample. The potassium hydroxide is good for approximately 250 test samples. The technician simply reads the fluid level on the scale, which is calibrated directly in percentage of CO_2. If the CO_2 reading is low, it is an indication that there is too much excess air, or that the fuel is not burning completely.

More common today are newer digital combustion analyzers like the one shown in Figure 2-8 on page 15. They are easy to use and offer many additional features. Such devices are capable of providing multiple readings and calculations from a single air sample. Be aware that the sensors have a limited life and must be maintained per the manufacturer's specifications to ensure accurate readings.

The first step in checking combustion is to make sure that the firing rate of the burner is correct. To check the firing rate of a gas burner, verify that the manifold pressure matches the manufacturer's requirements. The gas meter, when available, also should be clocked to determine the actual rate of gas usage (all other gas appliances must be turned off). For oil burners, check the oil pressure.

Basically, the *efficiency* of a furnace or boiler is the difference between the *possible* heat output

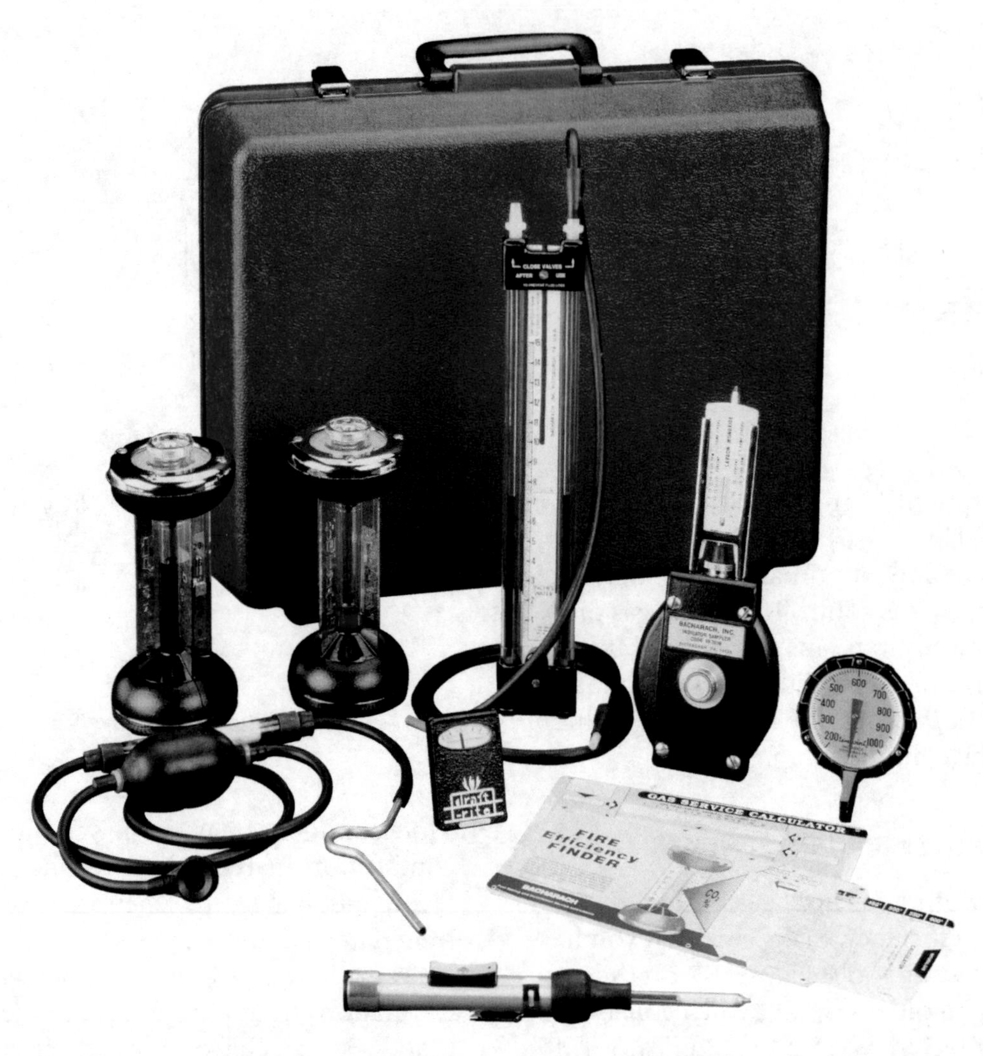

Figure 2-7. Gas burner combustion test kit

and the *actual* amount of heat obtained from the unit. A furnace will put *x*% of the heat obtainable from burning gas into the air as it passes through the furnace. A boiler will put *x*% of the heat obtainable from burning gas into the water as it passes through the boiler.

Combustion efficiency is an index of the useful heat obtained from the gas being burned, expressed as a percentage of the total heat produced by the burning of the gas (assuming complete combustion with no excess air). The percentage of CO_2 and the net temperature of the flue gases are the two measurements that establish combustion efficiency. These two measurements should be taken on the inlet side of the draft hood. The flue gas thermometer is inserted into the same hole used for the CO_2 test.

The overall objective of efficiency testing and adjustment is to obtain all the heat possible in the conditioned space from the potential amount in the gas without creating undesirable conditions. To do this, several factors are involved:

- overall combustion performance
- burner performance
- heat exchanger operation.

Use Figure 2-9 as an efficiency checklist. These factors are interrelated and therefore affect each other. A change in one will cause changes in all the others. When performing efficiency tests and making adjustments, make sure that you take all of them into account.

SEQUENCE OF OPERATION

Although there are many minor variations from manufacturer to manufacturer, the general sequence of operation is the same for all gas furnaces. There are three primary forms of ignition used with gas furnaces—the standing pilot system, intermittent ignition, and direct ignition.

Standing pilot

A *standing pilot* furnace provides a constant flame that is available at all times to light the main burner. The standing pilot system generally uses

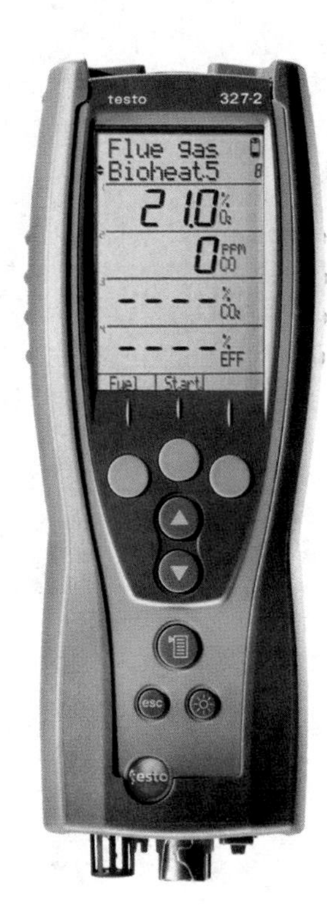

Figure 2-8. Digital combustion analyzer

Input

Type of gas ______________________

Btu content ______________________ per ft^2

Orifice drill size ______________________

Manifold pressure ______________________

Volume of gas used ______________________ ft^3

Flame appearance

Type of primary adjustment:

- ☐ Wing
- ☐ Butterfly
- ☐ Disk

Flame before adjustment:

- ☐ Sharp blue
- ☐ Red tips
- ☐ Heavy yellow

Neutral point adjustment

- ☐ Factory-designed (not adjustable)

Conversion burner:

Below adjustment range ______________________

Above adjustment range ______________________

Incorrect adjustment range ______________________

Air temperature rise

Supply air temperature ______________________

Return air temperature ______________________

Temperature rise ______________________

CO_2

First test: CO_2% ______________________

Second test: CO_2% ______________________

Final: CO_2% ______________________

Figure 2-9. Efficiency checklist

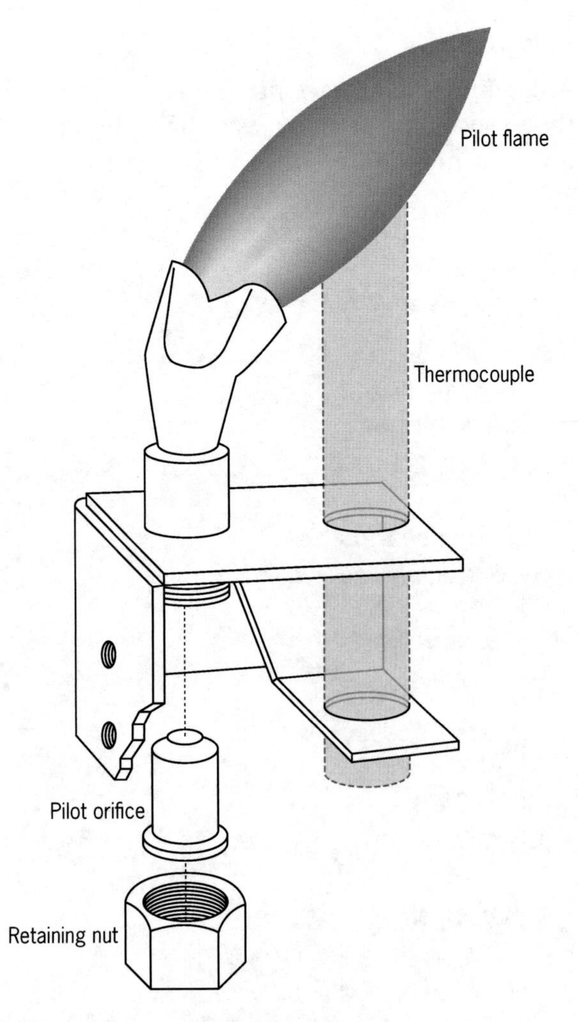

Figure 2-10. Standing pilot burner assembly with thermocouple

a *thermocouple* (see Figure 2-10) to provide proof that the pilot flame is present and available to light the main burner. In the presence of heat from the flame, the thermocouple produces a small voltage (approximately 18 to 30 mV). The voltage from the thermocouple holds the pilot valve open and allows gas to continue to flow to the pilot burner and to the secondary or main gas valve. If the pilot flame goes out, the thermocouple stops producing voltage, which causes the pilot valve to close. Once the pilot valve has closed, it also prevents gas from flowing to the main valve (thus keeping gas from flowing to the burners).

On a call for heat, the thermostat for a standing pilot furnace feeds 24 V from the "W" terminal to the main gas valve. The main valve is energized—and as long as the pilot valve is being held open by the thermocouple, gas will flow to the burners and be ignited by the pilot flame. After a brief warm-up period, a temperature-operated fan switch in the furnace closes and the blower starts. Once the call for heat has been satisfied, the gas valve is de-energized and the burner stops. After a cool-down period, the temperature-operated fan switch opens and the blower stops.

Intermittent ignition

An *intermittent ignition* system is one that lights a pilot on a call for heat and then, after the pilot flame has been proven, the main gas valve opens to light the burner. Most intermittent ignition systems use *spark igniters* (see Figure 2-11) to light the pilot and *flame rectification* (see Figure 2-12)

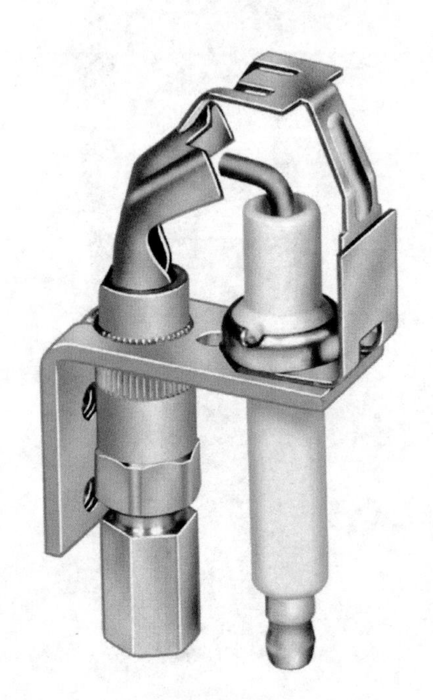

Figure 2-11. Intermittent pilot burner

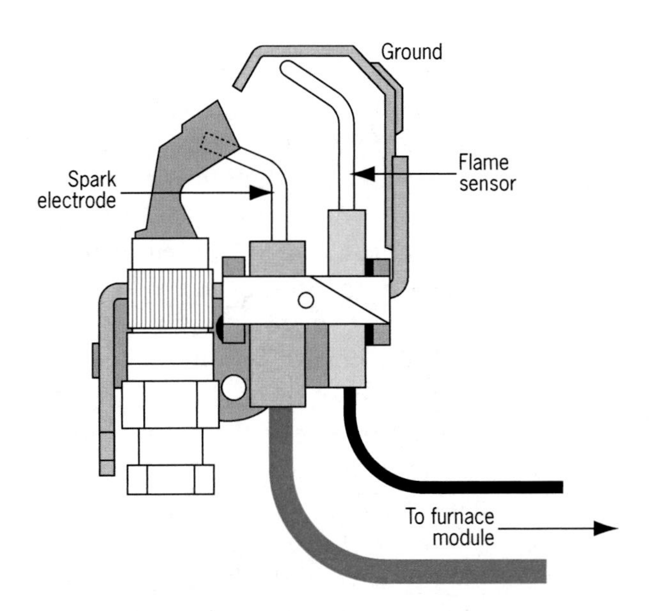

Figure 2-12. Flame rectification sensing system

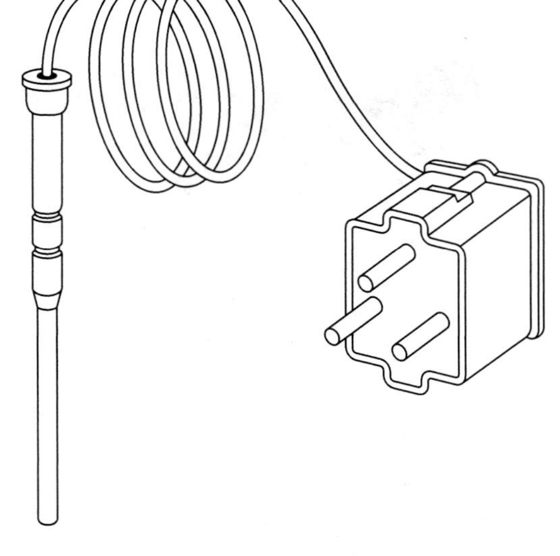

Figure 2-13. Mercury flame sensor plugs into gas valve

to prove the pilot. Some older systems use a mercury flame sensor (see Figure 2-13), which is a relatively quick-acting temperature-sensitive switch.

On a call for heat, the thermostat feeds 24 V to the ignition module. If the furnace has an induced-draft fan, the fan starts and a pressure differential switch closes to prove air flow in the vent. The module then begins the spark and opens the pilot gas valve. Once the pilot lights, an applied ac voltage passes through the flame—but only in one direction, thus providing a dc voltage back to the module via a rectification sensor (see Figure 2-14). As soon as the module sees the dc signal created by the presence of the flame, it opens the main valve to light the burner. After a short warm-up period, the module starts the blower. Once the call for heat has been satisfied, the gas valve closes and the burner stops. After a cool-down period, the module shuts down the blower. Manufacturers use various voltages for hot-surface igniters. Check the installation instructions for confirmation of the voltage used.

Direct ignition

A *direct ignition* system is one that lights the main burner directly and does not use a pilot light. Direct ignition systems commonly use *hot-surface igniters* (see Figure 2-15), but can also use spark igniters. These systems use the flame rectification method to prove that the main burner has lit. The flame rectification system provides a very fast response, which is critical with direct ignition.

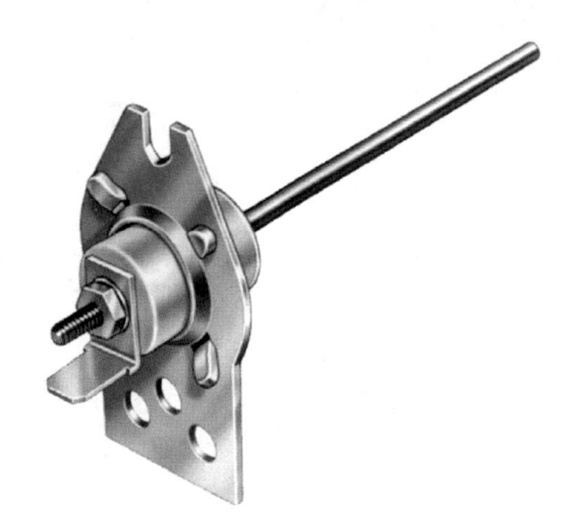

Figure 2-14. Typical rectification sensor

Figure 2-15. Hot-surface igniter

On a call for heat, the thermostat feeds 24 V to the ignition module. The induced-draft fan starts and a pressure differential switch closes to prove air flow in the vent. The module then energizes the hot-surface igniter or begins the spark. The hot-surface igniter is allowed to warm up for approximately 30 seconds and then the gas valve is energized. As gas flows to the main burner, it is ignited by the hot-surface igniter or the spark.

Once the burner lights, an applied ac voltage passes through the flame—again, only in one direction, thus providing a dc voltage (rectification) back to the module. If the module fails to see the dc signal created by the presence of the flame in a very short period of time, it shuts down the gas valve. If the flame is proven, the module starts the blower after a short warm-up period (based on time). Once the call for heat has been satisfied, the gas valve closes and the burner stops. After a cool-down period (based on time), the module shuts down the blower.

Figure 2-16. Combination gas control

GAS VALVES

Today's typical gas furnace uses a *combination gas valve*, as shown in Figure 2-16. The combination valve includes two separate gas valves—for the main gas and the pilot gas—plus a manual shutoff and a pressure regulator. In a standing pilot system, the pilot valve is operated by the millivolt signal produced by the thermocouple, and the main valve is opened by the 24-V signal from the thermostat. In intermittent ignition and direct ignition systems, both valves are controlled by the 24 V coming from the ignition module. Two-stage gas valves are also available. A two-stage gas valve includes an additional regulator, the first stage of which operates at roughly half the pressure of the second stage. □

REVIEW QUESTIONS

1. Which of the following is required for combustion?

   a. Fuel
   b. Heat
   c. Oxygen
   d. All of the above

2. Which of the following is *not* a product of complete combustion?

   a. Carbon dioxide
   b. Carbon monoxide
   c. Heat
   d. Water

3. An *unconfined space* (in terms of furnace installation) is defined as a space with a volume equal to or greater than _________ per 1,000 Btu of input rating.

   a. 25 ft^3
   b. 50 ft^3
   c. 75 ft^3
   d. 100 ft^3

4. Combustion air may be added to a confined space by a duct to the outside, sized on an input of 1 in^2 per _________.

   a. 1,000 Btu
   b. 2,000 Btu
   c. 4,000 Btu
   d. 5,000 Btu

5. What type of burners are commonly used with induced-draft gas furnaces?

   a. Gun style
   b. Pot style
   c. Ribbon
   d. Inshot

6. What is the typical color of a properly burning gas flame?

   a. Blue
   b. Orange
   c. Red
   d. Yellow

7. In a gas furnace, primary combustion air is introduced in the _________.

   a. burner prior to ignition
   b. burner after ignition
   c. gas line
   d. heat exchanger

8. A "dumbbell" style CO_2 tester _________.

   a. contains a test fluid that absorbs CO_2
   b. measures CO_2 in parts per million
   c. provides a digital readout
   d. requires a power supply

9. Useful heat output, expressed as a percentage of the total burner input, is known as the _________ of a gas furnace.

   a. combustion efficiency
   b. fuel efficiency
   c. heat exchanger performance factor
   d. heating coefficient

10. Which type of ignition system uses a thermocouple?

    a. Direct ignition
    b. Hot-surface
    c. Intermittent ignition
    d. Standing pilot

11. The blower in a standing pilot furnace is generally started by a(n) _________.

    a. differential pressure switch
    b. ignition module
    c. temperature switch
    d. time-delay relay

12. A furnace with an induced-draft fan uses a _________ to prove air flow.

    a. high-pressure switch
    b. low-pressure switch
    c. pressure differential switch
    d. sail switch

13. *Flame rectification* proves flame by creating a signal in response to a _________.

    a. change in resistance
    b. current conducted in one direction through the flame
    c. temperature-activated switch
    d. voltage produced by the flame

14. Which of the following is contained within a combination gas valve?

    a. Main gas valve
    b. Manual shutoff valve
    c. Pressure regulator
    d. All of the above

15. The gas valve in a direct ignition system is operated by a _________.

    a. 24-V signal from the ignition module
    b. control voltage from the temperature switch
    c. dc voltage from the flame rectifier
    d. small voltage from the thermocouple

ILLUSTRATION CREDITS
FIGURE 2-3: ILLUSTRATED HOME BY CARSON DUNLOP COMPANY
FIGURE 2-7: BACHARACH, INC.
FIGURE 2-8: TESTO, INC.
FIGURE 2-11: HONEYWELL
FIGURE 2-14: HONEYWELL
FIGURE 2-15: WHITE-RODGERS
FIGURE 2-16: HONEYWELL

CHAPTER 3

Furnace Installation

INTRODUCTION

The first step in any furnace installation is to select the proper equipment for the job. Equipment selection must start with a proper load calculation for the space to be heated. The load calculation should be done in accordance with ACCA's *Manual J*. A load calculation takes into consideration the different types of construction materials used in a building, the outside design temperature for the location where the unit is to be installed, and the design inside temperature for the space to be conditioned. These factors are used in determining the temperature difference needed to calculate the heat loss through walls, windows, doors, ceilings, and floors. A further calculation is done to account for ventilation or air leakage into the conditioned space. When a load calculation is done for cooling, heat gain from the sun and internal loads such as lights, people, and equipment also need to be added.

The furnace should be sized based on the Btu output that most closely matches the calculated heat loss. A furnace that is undersized will not be able to maintain the desired temperature, and one that is oversized will *short cycle* (turn off and on frequently), which reduces the efficiency and shortens the overall life of the equipment. When the furnace output matches the heat loss of the space, the system is said to be at its *balance point* and the furnace will run 100% of the time to maintain the design temperature in the space.

LOCATIONS

The location of a furnace generally is dictated to some extent by the construction of the building. Since floor space is at a premium, the furnace is often located in a basement, crawl space, attic, garage, or utility closet. Some of these locations are unheated, which means that steps must be taken to protect condensate lines and water lines for humidifiers from freezing. It is also important to insulate and provide a vapor barrier for any ductwork that runs through unconditioned spaces.

Basements

Most basement installations use upflow units, with discharge through the floors overhead (see Figure 3-1). The typical basement has a concrete floor, which provides a substantial,

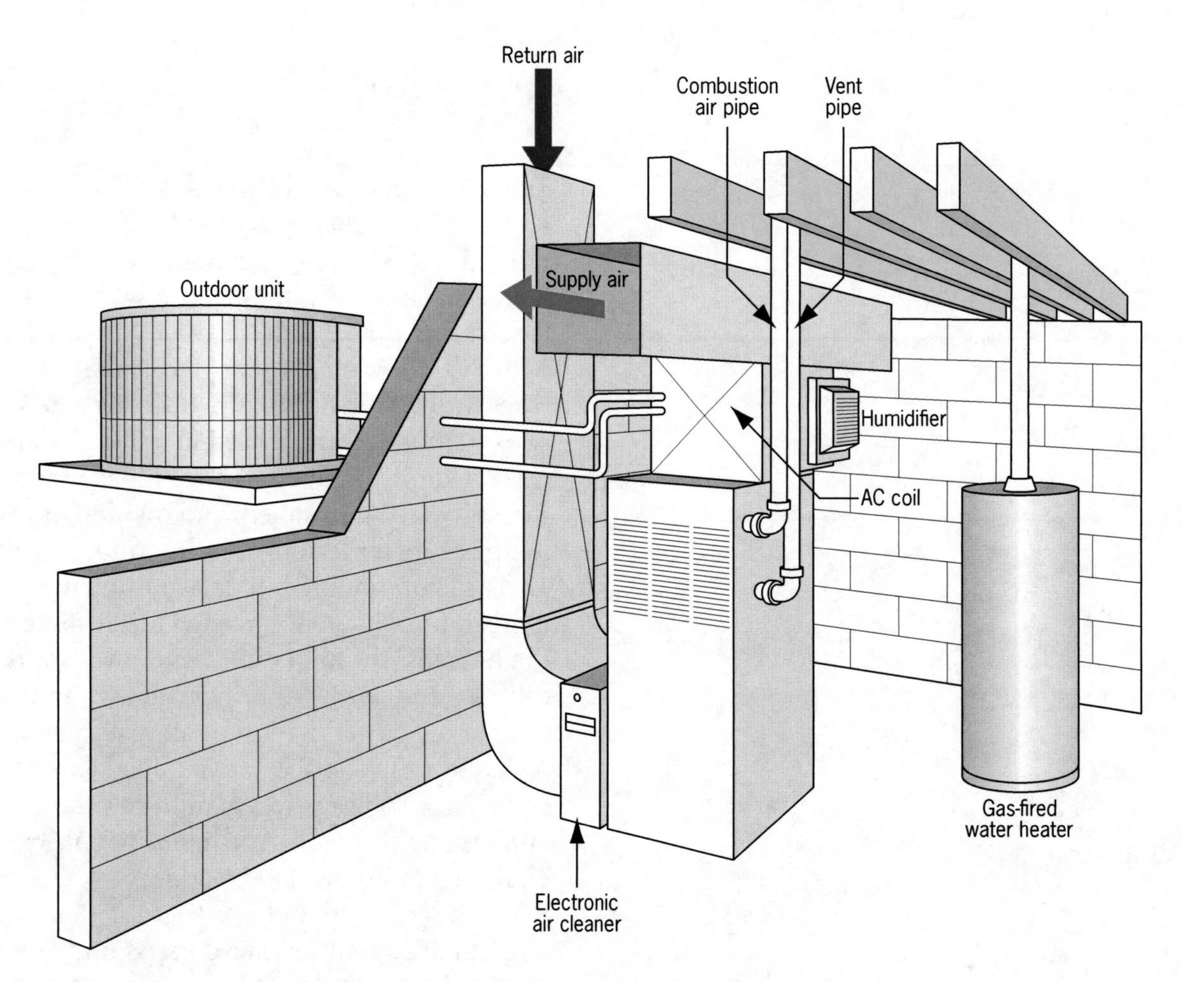

Figure 3-1. Basement installation of upflow gas furnace

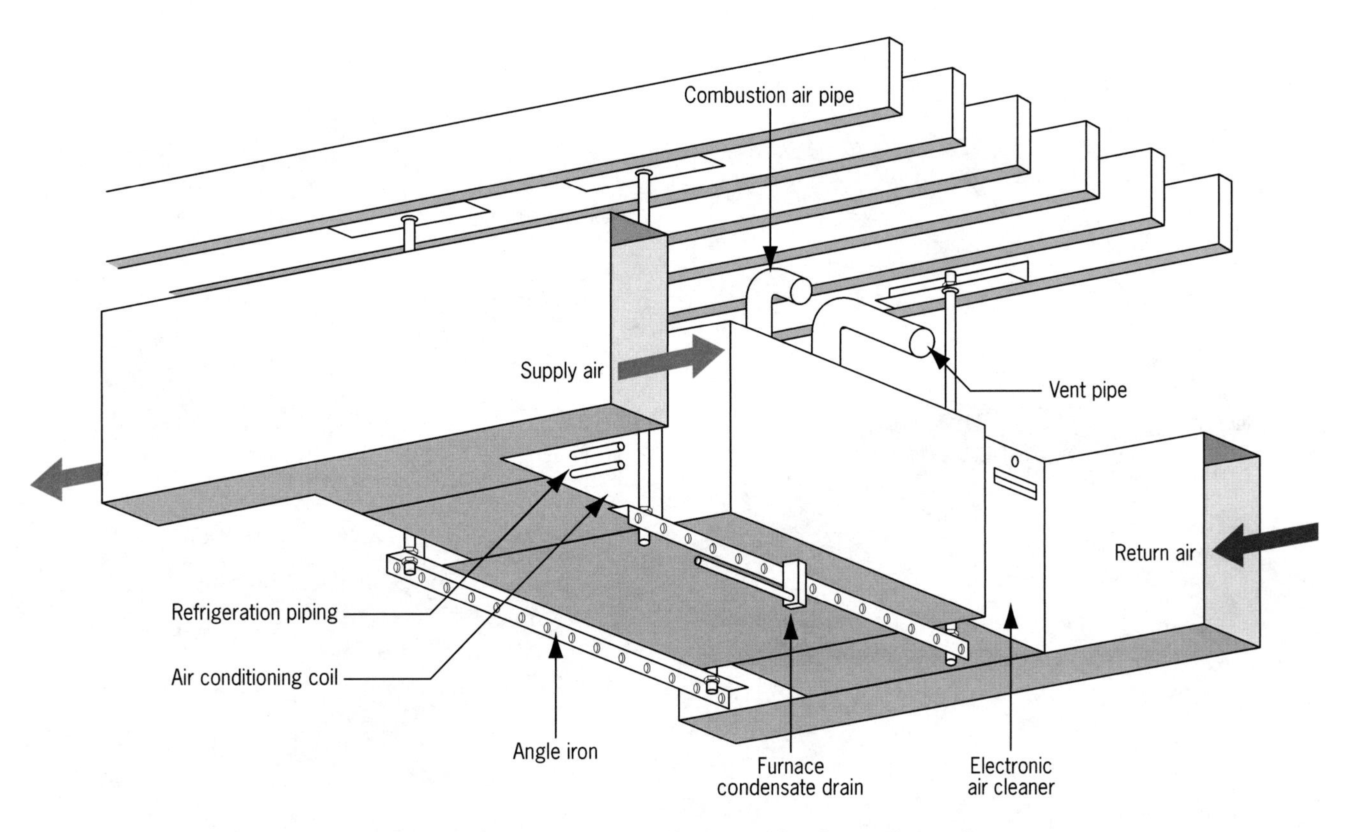

Figure 3-2. Crawl space installation of horizontal gas furnace

non-combustible base. A basement installation requires the designer and/or installer to consider the location of fuel lines, combustion air and venting, the condensate drain, and duct runs.

Crawl spaces

Crawl spaces usually do not have finished floors, which means that the unit frequently is hung horizontally from the floor joists, as shown in Figure 3-2. The ductwork must maintain a minimum clearance above the ground to prevent moisture infiltration. The duct also needs to be insulated and covered with a vapor barrier to prevent unnecessary heat loss.

Attics

Attic installations frequently call for a horizontal unit to be hung from the roof rafters. In some cases, the unit sits on the ceiling joists, as shown in Figure 3-3 on the next page. Since truss-type construction is common today, it is critical for the load to be distributed evenly over several trusses. A base somewhat larger than the unit shall be provided, including a work platform in front of the unit to allow for safe service in the future. If the unit is mounted in a downflow configuration or a zero side clearance is not available, a non-combustible floor may need to be added.

Closets

Closet or utility room installations often are located near the center of the house. This location allows for a central return through the wall, thus eliminating the return duct system. While this arrangement reduces installation costs, it does not always provide the best air distribution. Supply ducts can be run overhead in the attic or underneath in the slab or crawl space. If the unit

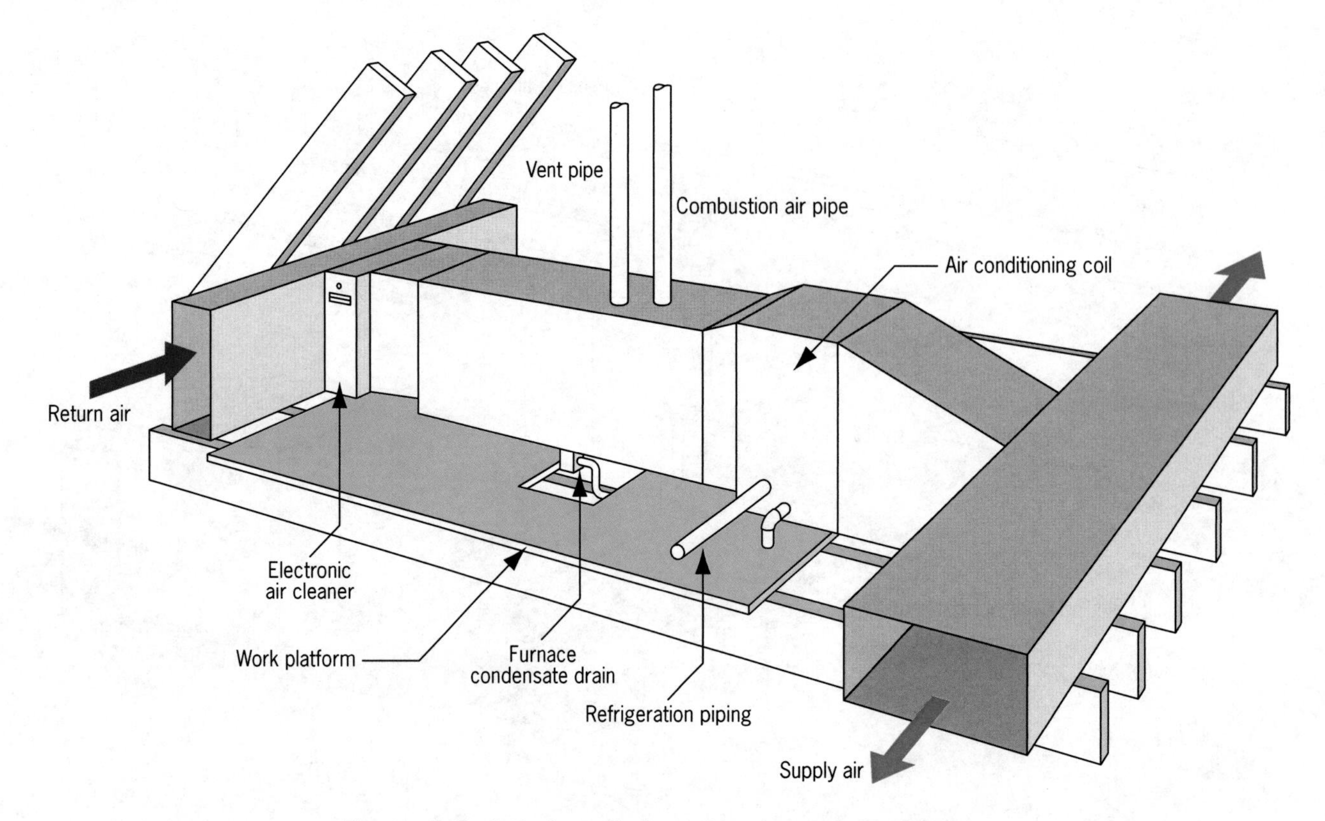

Figure 3-3. Attic installation of horizontal gas furnace

is being used in a downflow configuration, a noncombustible floor may need to be installed.

Garages

A furnace installed in a garage must be mounted off the floor (to reduce the risk of igniting gas fumes that tend to settle near the floor). How far off the floor can vary—check with the authority having jurisdiction (AHJ). Ducts within the garage space must be sealed in order to prevent fumes in the garage from entering the duct system and being carried into the house. There may also be requirements for fire dampers where the ductwork passes between the house and garage.

ELECTRICAL CONNECTIONS

A furnace generally operates on a 120-V dedicated circuit. In many localities, the circuit must be installed to the equipment disconnect by a licensed electrician. The connection from the disconnect to the equipment should be checked by the HVAC technician for polarity. In all cases, the technician should verify that all connections have been properly made and that all terminals have been tightened down. Most jurisdictions follow the NEC (National Electrical Code) with local amendments. It is important to check with the proper authority in your area.

In addition to the line-voltage connection, the low-voltage connection between the furnace and the thermostat must be installed. A multiconductor cable is generally run from the furnace to the thermostat for this purpose. A separation must be maintained between the low-voltage cable and the line-voltage wiring—per the NEC, they *cannot* share the same conduits.

The wiring connections to the thermostat are made at the subbase. The most common terminal designations on thermostats are R, G, Y, W, and (on some electronic thermostats) C. These terminals are connected to the corresponding terminals on the furnace terminal strip or ignition control board. Additional connections may be required for some furnaces, depending on what options and accessories are used.

FUEL PIPING

A connection must be made from the fuel supply to the furnace. Gas generally is supplied to the building at a pressure of no more than 0.5 psi, or about 14 in. w.c. Pressures above this level can damage the gas valves at the equipment. Due to the very low pressure, the piping system must be sized to provide the volume of gas needed by the equipment with a very minimal pressure drop. It is important to follow the requirements of the National Fuel Gas Code (NFPA 54).

Natural gas systems generally use threaded steel pipe or corrugated stainless steel tubing (CSST) to supply the gas to the furnace (chemicals within some gas supplies can damage copper pipe). Pipe thread sealant should be applied on the male thread only. To prevent the pipe dope from being pushed inside the pipe during assembly, avoid getting any of the sealant on the very end thread. If Teflon® tape is to be used, make sure to utilize only the type that is rated for use with natural gas. A sediment trap and union should be provided at each unit. The sediment trap catches and prevents moisture, oil, and particles from contaminating the gas valve. The union provides easy disassembly for service. Figure 3-4 shows a typical gas piping layout.

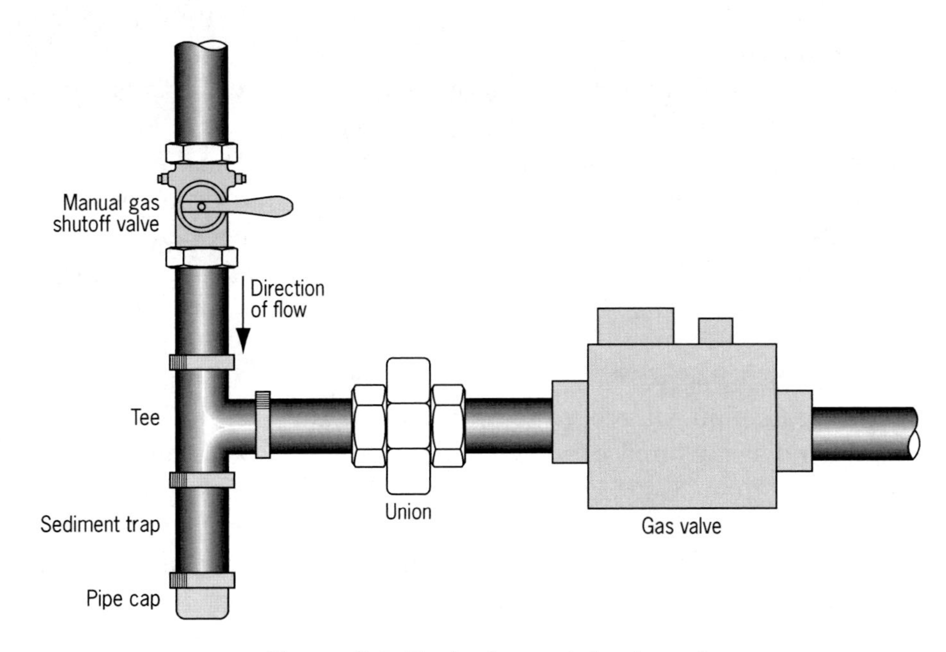

Figure 3-4. Typical gas piping layout

In oil heating systems, piping between the oil tank and the oil pump usually takes the form of copper tubing with flared connections. More details on oil tanks and piping can be found in Chapter 8.

Gas leak detection

Once all of the piping has been assembled, it is important to check the piping for leaks. Use a combustible gas detector, not a refrigerant detector (a refrigerant detector will not detect combustible gases). An electronic detector like the one shown in Figure 3-5 on the next page is sensitive to gas at levels as low as 20 to 50 ppm (parts per million). A typical electronic detector has an audible alarm that gets louder or faster

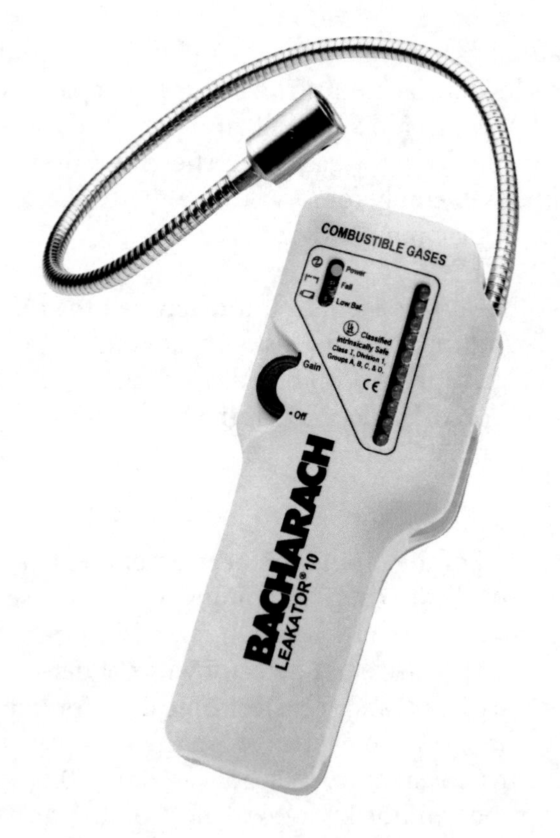

Figure 3-5. Combustible gas leak detector

as the concentration of gas increases. Do *not* use an electronic leak detector in locations where moisture can enter the tip and damage the probe. *Note:* It is important to check the leak detector near the pipe sealant to see if there is a reaction. Some pipe sealants use solvents that give off gases that will be detected by the instrument. Ultrasonic leak detectors and soap bubbles also can be used.

Condensate piping

Condensate piping is needed to carry condensate from condensing furnaces and air conditioning coils to the drain. Condensate piping should have a nominal diameter of at least 3/4 in. and should be sloped toward the drain at a minimum pitch of 1/8 in. per foot. Condensate from the air conditioning coil and the furnace can share the same line, but each should have its own trap. The drain line should be open-vented. *Never* pipe the drain line solidly into the sewer. If you do and the sewer line becomes blocked, backups into the furnace can occur.

PVC pipe generally is used for condensate piping. The material is lightweight and easy to work with. It should be cut and dry-fitted prior to gluing. When gluing together a pipe and fitting, apply the appropriate primer to both surfaces first, then apply the glue. Rotate the pipe slightly during assembly to ensure even coverage of the glue. The joint should be held in place for a short time to allow the glue to tack.

THERMOSTATS

The *thermostat* is the main operating control for a furnace. It tells the furnace to start and stop as needed to maintain the desired temperature within the space. A thermostat is basically a temperature-activated switch. When the temperature within the space drops below the setpoint, the switch closes and feeds 24 V back to the furnace on the "W" terminal to start the furnace. Once the setpoint has been reached, the switch reopens and the furnace shuts down.

The thermostat should be installed in a central location where it will be subject to the "average" temperature of the conditioned space. This means that the thermostat should be mounted approximately 5 ft above the floor on an interior wall. Other guidelines for installing thermostats include:

- Do not mount the thermostat where it will be exposed to direct sunlight.
- Do not mount near supply registers or other sources of heat.
- Do not mount near outside doors or on an exterior wall.
- Do not mount in or near the kitchen.

A basic mechanical thermostat is simply a set of switches. It does not need power to operate, and therefore does not have a common connection from the 24-V power supply. If an electronic thermostat or a mechanical thermostat with lights is to be installed, then a common connection may be required to provide power. Typical connections for today's electronic thermostats are listed below:

Terminal	Function
R	24-V power
G	Fan circuit
Y	Cooling circuit
W	Heating circuit
C	24-V common

If the fossil fuel furnace is being used with a heat pump, the furnace will be connected to the W2 (second-stage heat) terminal. This allows the heat pump to operate until it can no longer keep up with the heat loss of the space. Additional controls are required to prevent the fossil fuel furnace from running while the heat pump is operating. On a call for second-stage heat, the heat pump will shut down and the furnace will start. When the second stage is satisfied, the controls will give the furnace time to cool down before allowing the heat pump to restart.

Electronic thermostats come in a variety of configurations, from the very simple to those that can accommodate multiple programs per day. A programmable thermostat allows for automatic temperature setback—thus saving energy when residents are sleeping or away. Some thermostats also have an *auto changeover* function that switches the system operating mode from heating to cooling, and vice versa, without a manual control. The difference between the actual heating and cooling setpoints is known as *deadband*. Since no action takes place in the deadband region, it prevents the equipment from cycling back and forth between heating and cooling when there is a slight overshoot in temperature. Figure 3-6 shows a programmable thermostat.

ACCESSORIES

Air conditioning

Air conditioning is often added to a heating system to provide cooling during the summer months. It is important to remember that when a cooling coil is added to the furnace, it should always be mounted on the *downstream* side of the heat exchanger. If the cold air from the coil were to blow directly onto the heat exchanger, the warm, moist air on the combustion side could cause condensation on the cold metal surfaces. Over time, such condensation would cause rusting and premature failure of the heat exchanger. Consult the *Air Conditioning and Heat Pumps* book in this series for more information about air conditioning systems.

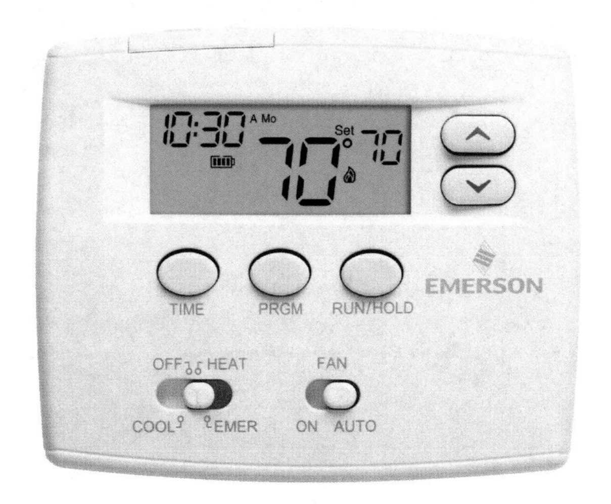

Figure 3-6. Programmable thermostat

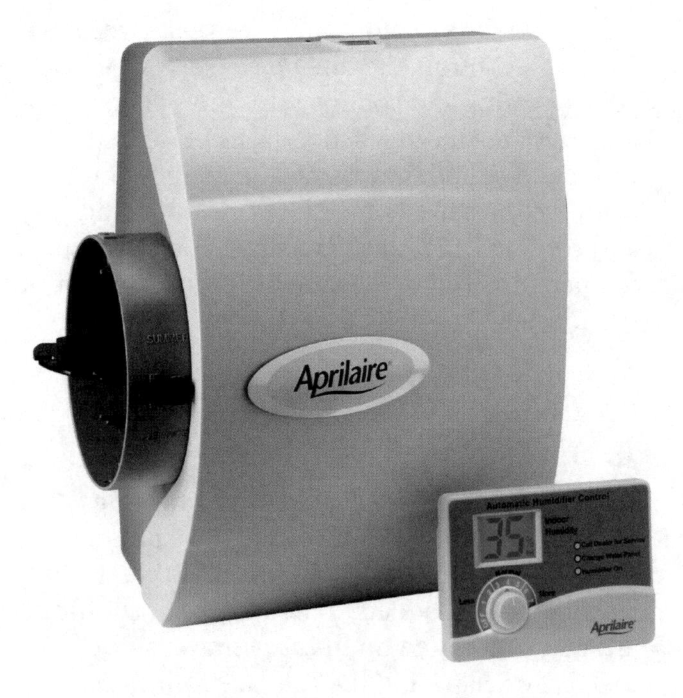

Figure 3-7. Bypass humidifier

Figure 3-8. Fan-powered humidifier

Humidifiers

Because the human body is cooled by the evaporation of moisture from the skin, and because the relative humidity affects how fast moisture will evaporate, it is beneficial to control the humidity within a conditioned space. By using a *humidifier* to increase the relative humidity in a room, the evaporation of moisture from the skin can be slowed—and, as a result, the occupant of the room will feel warmer even without increasing the temperature. Increasing the relative humidity also eliminates the static electricity that is often present with dry air.

Some people believe that the furnace "dries out" the air in the winter. In reality, it's simply a matter of cold air having far less capacity to hold moisture than warm air. Since cold air from outside is being used to ventilate the space (either through leakage or some form of controlled air intake), when it warms up it will still contain the same amount of moisture—but *its ability to hold moisture will increase*. And that means that its relative humidity actually *decreases*.

Humidifiers are sized based on the total cubic feet of space being heated. The outside temperature and the desired relative humidity also must be considered. These factors determine the amount of moisture that will need to be added to the conditioned space. If the relative humidity is too high for the outside temperature, condensation will form on cold surfaces (such as windows). Humidifiers are available that can make an adjustment to relative humidity based on the outside temperature.

There are several types of humidifiers that can be attached to a furnace to increase the relative humidity in the conditioned space. Two of the most common are bypass humidifiers and fan-powered humidifiers. A *bypass humidifier* (see Figure 3-7) is connected between the supply side of the furnace and the return side. The pressure difference between the warm supply air and the cool return air causes some air to flow from the supply duct, through a wetted pad inside the humidifier, and into the return duct. A manual damper is generally in place in the bypass duct. This damper should be closed in the summer to prevent the bypass of air during the cooling season. A *fan-powered humidifier* (see Figure 3-8) has its own fan, which draws air from the duct, pushes it through a wetted pad, and discharges it back into the same duct.

Figure 3-9. Steam humidification system

Steam and atomization humidification systems are also available. *Steam* units like the one shown in Figure 3-9 generate steam and inject it directly into the supply airstream. *Atomization* units generate a very fine mist of water droplets and inject it by means of a special nozzle directly into the return airstream. Figure 3-10 shows an atomizing humidifier.

ZONING

Zoning provides multiple temperature-controlled areas within a single structure. These individual areas ("zones") may be separate rooms, or groups of rooms, each of which has its temperature controlled by a single thermostat. Air flow to different zones may be controlled by individual zone dampers, or zones may be heated and cooled by separate equipment.

Figure 3-10. Atomizing humidifier

When zone dampers are used to provide temperature control, technicians must take special precautions to ensure that proper air flow is maintained through the heating and cooling equipment at all times. This includes times when the smallest system zone is calling. During such times, proper air flow often is maintained by using a *bypass damper*. Bypass

Figure 3-11. Bypass damper

dampers may be motorized or simple dampers with counterweights (barometric) that allow air flow to pass when the system static pressure exceeds a certain point. Figure 3-11 shows a typical bypass damper.

One technique frequently used in zoning systems is to mount a bypass damper in a duct running from the supply to the return. Excess air is taken from the supply side of the air distribution system and routed ("bypassed") directly into the return airstream. Installers must be careful to tap into the return ductwork as far away as possible from the equipment. This allows the return air to mix with the conditioned supply air before re-entering the heating or cooling equipment. If the bypass connection is not located properly, system problems can result.

In the heating mode, directing the heated supply air directly back into the return air may cause overheating of the furnace and subsequent limit trips. A high return air temperature can cause electronic components within the furnace (such as control boards) to overheat. Most equipment manufacturers specify a maximum recommended return air temperature, which you normally will find printed on the furnace rating plate. In the cooling mode, cool conditioned air bypassed directly into the return air may cause the evaporator coil to "freeze," a situation that can lead to liquid refrigerant floodback to the compressor and eventual compressor failure.

A common alternative to bypassing the supply air directly to the return is to employ a *dump damper*. This practice works well in installations

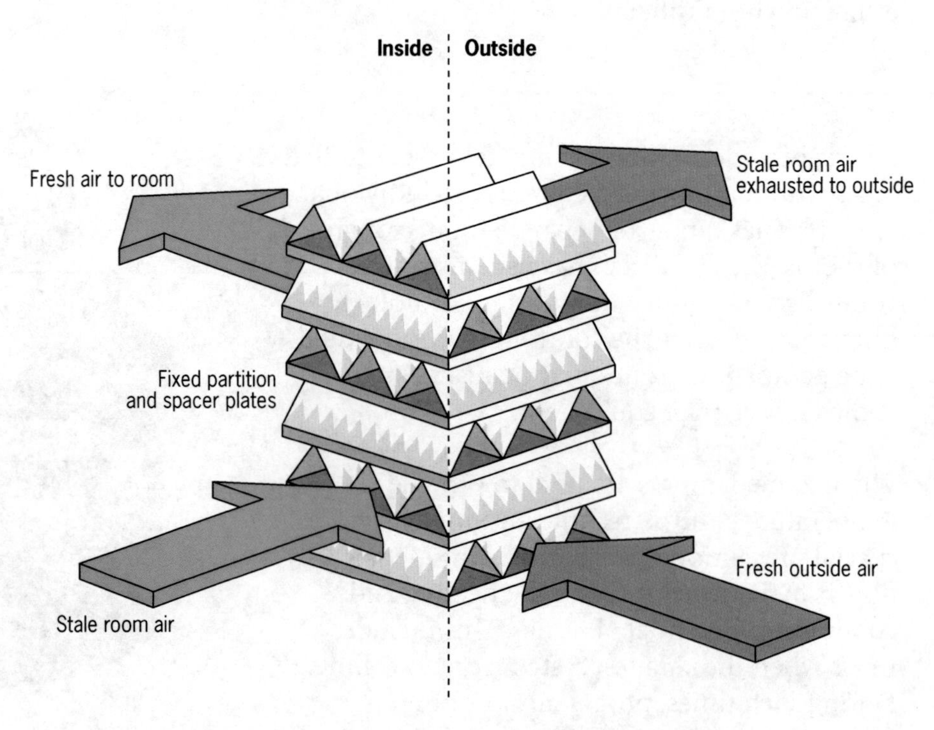

Figure 3-12. Heat recovery ventilator (HRV)

that have areas where temperature is not critical, such as basements or crawl spaces. A barometric damper is used, and the counterweight is adjusted by the installing technician to open only when the call is coming from the smallest system zone(s). The bypass damper is not routed to the return duct—instead, it "dumps" the excess pressure into an unoccupied area of the home. Installation should be in accordance with the ACCA/ANSI zoning standard.

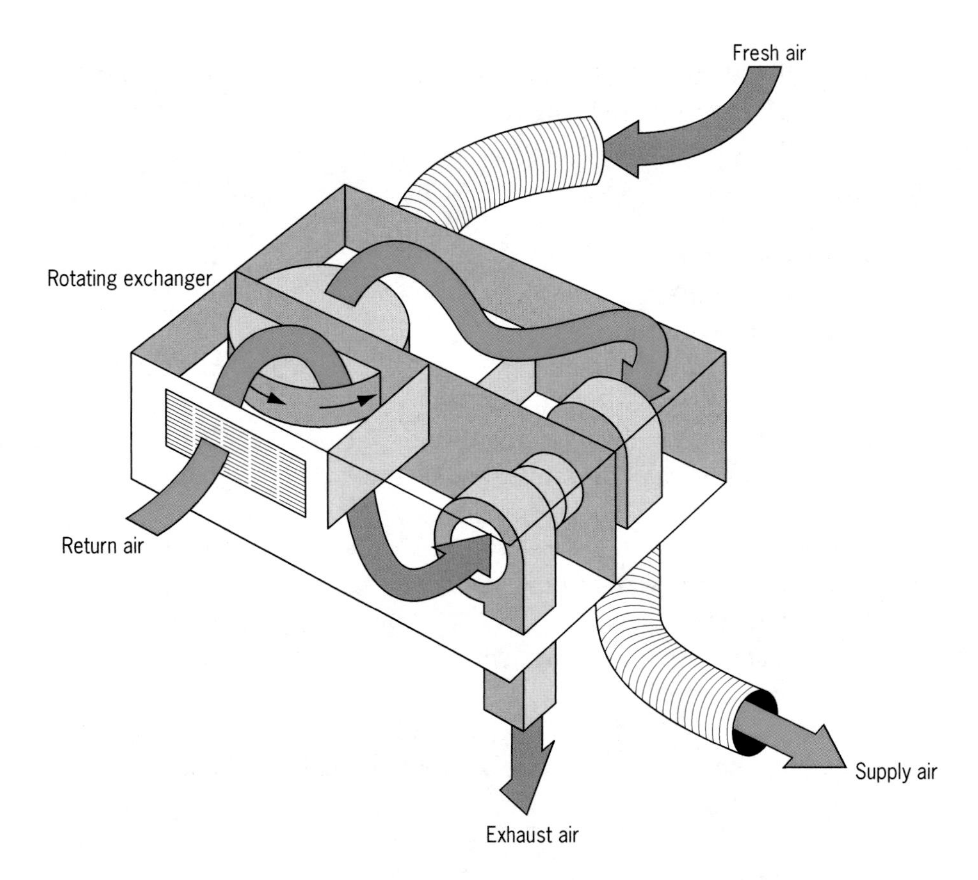

Figure 3-13. Energy recovery ventilator (ERV)

FRESH AIR INTAKES AND ECONOMIZERS

Some systems may add *fresh air intakes* to provide ventilation to the conditioned space. *Economizers* are sometimes referred to as "free" cooling systems. An economizer acts as a part of the fresh air intake and is installed when the system requires cooling even in cooler weather. The economizer opens and tries to cool the space with the cooler outside air before it starts mechanical cooling. When installing these units, make sure that they are aligned properly and that the damper blades move freely.

More and more applications are using *heat recovery ventilators* (HRVs) and *energy recovery ventilators* (ERVs) to provide some fresh air for ventilation while extracting the heat energy from the air being exhausted. Figure 3-12 shows how an HRV works. The air being exhausted and the air being brought in from outside pass through a heat exchange core where sensible heat is transferred from the outgoing airstream to the incoming airstream. Figure 3-13 shows an ERV with a rotating heat exchanger where both latent and sensible heat transfer take place. Requirements regarding the installation of HRVs and ERVs are set forth in ASHRAE Standards 62.1 and 62.2. □

REVIEW QUESTIONS

1. Residential load calculations should be completed according to ACCA's __________.

   a. *Manual D*
   b. *Manual J*
   c. *Manual N*
   d. *Manual P*

2. When the output of a furnace matches the heat loss or heat gain of the conditioned space, the system is said to be at its __________.

   a. balance point
   b. design temperature
   c. full-load condition
   d. reference point

3. When furnaces are installed in attics or crawl spaces, special care must be taken to __________.

   a. distribute the weight properly
   b. insulate the ductwork
   c. protect water lines from freezing
   d. all of the above

4. Which of the following is true of low-voltage control wiring?

   a. It is exempt from NEC requirements
   b. It may not be run in the same conduit as line-voltage wiring
   c. It may not exceed 100 ft in length
   d. It must be run in metal conduit

5. What is the maximum pressure that can be supplied to most natural gas furnaces?

   a. 3.5 in. w.c.
   b. 14 in. w.c.
   c. 2 psi
   d. 5 psi

6. The sensor of an electronic leak detector can be affected by __________, which may cause the instrument to give a false indication of a gas leak.

   a. cutting oil
   b. moisture
   c. pipe sealant
   d. Teflon tape

7. What is the minimum diameter for condensate drain piping?

   a. $\frac{1}{2}$ in.
   b. $\frac{3}{4}$ in.
   c. 1 in.
   d. $1\frac{1}{4}$ in.

8. Generally speaking, a residential thermostat should be mounted __________.

   a. approximately 5 ft off the floor
   b. in the kitchen
   c. near a supply register
   d. on an exterior wall

9. Which is the proper terminal for heating on most thermostats?

   a. C
   b. G
   c. W
   d. Y

10. On an *auto changeover* thermostat, the difference between the heating and cooling setpoints (that keeps the equipment from cycling back and forth between heating and cooling) is known as the _________.

    a. deadband
    b. droop
    c. offset
    d. throttling range

11. In order to prevent condensation from forming in the heat exchanger of a furnace, the cooling coil should always be mounted _________.

    a. before the heat exchanger
    b. below the heat exchanger
    c. downstream from the heat exchanger
    d. upstream from the heat exchanger

12. In the summer months, the damper on a bypass humidifier should be _________.

    a. fully closed
    b. fully open
    c. partially open
    d. removed

13. Which of the following is *not* true of a residential zoned system?

    a. Each zone has its own temperature control
    b. There can be no more than four zones within a single structure
    c. Zoned units may use a bypass damper to pass excess supply air to the return duct
    d. Zoned units may use a dump damper to pass excess supply air to a non-critical area

14. A type of equipment that transfers both latent and sensible heat from the exhaust air to the supply air is known as a(n) _________.

    a. economizer
    b. energy recovery ventilator
    c. fresh air intake
    d. heat recovery ventilator

ILLUSTRATION CREDITS
FIGURE 3-1: CARRIER CORPORATION
FIGURE 3-2: CARRIER CORPORATION
FIGURE 3-3: CARRIER CORPORATION
FIGURE 3-5: BACHARACH, INC.
FIGURE 3-6: EMERSON CLIMATE TECHNOLOGIES, INC.
FIGURE 3-7: APRILAIRE
FIGURE 3-8: HONEYWELL
FIGURE 3-9: HONEYWELL
FIGURE 3-10: TRION, INC.
FIGURE 3-11: HONEYWELL
FIGURE 3-12: RENEWAIRE

CHAPTER 4

Venting

Editor's Note:
Venting information contained in this Chapter is for instructive purposes only. Always refer directly to the relevant codes for all requirements relating to proper venting procedures.

INTRODUCTION

Proper venting of fossil fuel furnaces is extremely important to the proper operation of the equipment and to the safety of the building's occupants. An improperly vented system may reduce the unit's efficiency or prevent the unit from operating. Leaking combustion products—which may include carbon monoxide—can cause severe illness or even death.

It is not the intent of this Chapter to provide a full guide to venting, but rather to highlight the basic procedures that can help the technician prevent costly or possibly even deadly mistakes. All systems must be vented in accordance with the manufacturer's specifications and all local codes. *The technician must read and follow the manufacturer's instructions for the equipment being installed.*

National Fuel Gas Code 12.7.3.3
"The sizing of gas vents for Category II, III, and IV appliances shall be in accordance with the appliance manufacturer's instructions."

DEFINITION OF GAS APPLIANCE CATEGORIES

Vented gas appliances are classified into four categories, defined by NFPA 54/ANSI Z223.1 as follows:

Category I

"An appliance that operates with a non-positive vent static pressure and with a vent gas temperature that avoids excessive condensate production in the vent."

Category II

"An appliance that operates with a non-positive vent static pressure and with a vent gas temperature that can cause excessive condensate production in the vent."

Category III

"An appliance that operates with a positive vent static pressure and with a vent gas temperature that avoids excessive condensate production in the vent."

Category IV

"An appliance that operates with a positive vent static pressure and with a vent gas temperature that can cause excessive condensate production in the vent."

Remember that the preceding definitions apply to the appliance and do not necessarily reflect the performance of the connected venting system.

Revisions to the standards now require the manufacturer to specify on the rating plate the category of the particular equipment—i.e., "Category I," "Category II," "Category III," or "Category IV." In addition, Category II, III, and IV equipment must bear a marking that states:

"This appliance requires a special venting system. Refer to installation instructions No. _____ for parts list and method of installation."

VENTING OF GAS-BURNING EQUIPMENT

When 1 ft^3 of natural gas is burned with just enough air for complete combustion, it produces 11 ft^3 of combustion products (see Figure 4-1). This total includes 1 ft^3 of carbon dioxide and 2 ft^3 of water vapor formed in the burning of the gas. It also includes about 8 ft^3 of nitrogen, which did not take part in the burning of the gas. In virtually all burners, some excess air is used. This volume of excess air is added to the combustion products flowing from the flue outlet. A provision is made on most equipment to vent (remove) these flue gases from the home.

There are basically two types of venting—power venting and atmospheric (or gravity) venting. *Power venting* makes use of a mechanical device, such as a blower, to draw flue gases from the area being vented. In *atmospheric venting,* hot flue gases from a burner pass into a vent connector, vent, chimney, or stack. The two types of venting systems perform the same function, but differ in the way they are constructed. Essentially, the term *gravity venting* as it is referred to in this Chapter designates a nonpowered venting system.

The driving force for a gravity vent comes from hot gases that tend to rise in the surrounding cooler air. Venting force is directly related to two factors—the density of the hot gases in the vent (which depends on the average gas temperature in the vent), and the height of the gravity vent. The hotter the gases in the vent and the higher

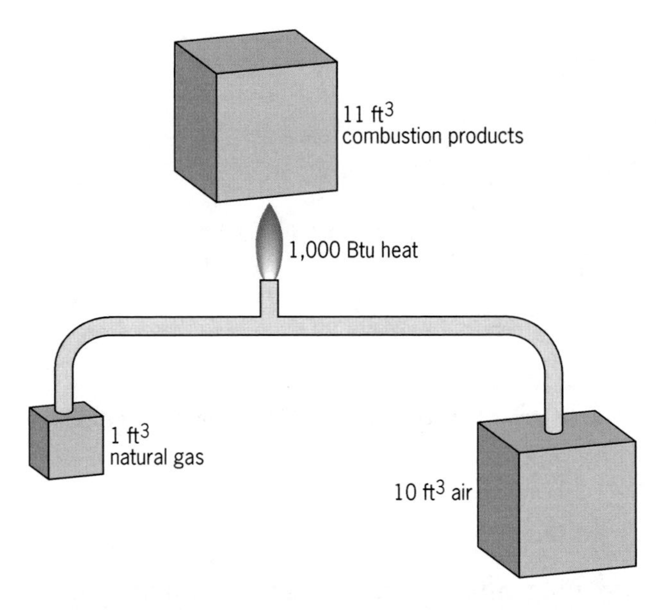

Figure 4-1. Combustion products from burning 1 ft^3 of natural gas

the vent pipe, the greater the venting force. The greater the venting force, the greater the amount of secondary air that will flow through the combustion zone.

It is important to understand this basic principle of gravity venting and its effects on the flow of air into the combustion zone. Its practical use lies in dealing with burner problems stemming from poor venting conditions in the combustion air zone (CAZ).

The flow of air into and combustion products out of the combustion zone must be reasonably close to the volume for which the unit was designed. More importantly, sufficient air must be provided to ensure complete combustion. On the other hand, too much air can reduce heating efficiency. Flue losses tend to increase as the volume of hot flue products increases.

A gravity vent must be connected to the flue outlet through a draft hood—not directly to the appliance. A direct connection will cause the amount of combustion, ventilation, and dilution air entering the combustion chamber to vary in relation to the height of the vent and the atmospheric conditions (e.g., wind). As a result, there will be little or no chance of maintaining the same air flow rate through an appliance for the variety of installation conditions that might be encountered.

In addition, gravity vents are exposed to outdoor winds. These winds can impose pressures at the terminal (exhaust end) of the vent, disturbing the natural draft. If wind creates a negative pressure at the vent terminal, it will tend to increase the flow out of the terminal—referred to as an *updraft*. If a wind creates a positive pressure at the vent terminal, it will tend to restrict flow from the vent—referred to as a *downdraft* or *backdraft*.

The exception to these comments on gravity vents is the *sealed combustion* type of appliance. In these units, inlet air is drawn from the outside as close as possible to the same points where the flue exhaust leaves the vent terminal. Wind can act on both the air inlet and the exhaust outlet in the same pressure plane at the same time, so that their effects cancel each other out. Balanced flue appliances, if properly designed, are largely immune to wind effects.

To overcome wind effects, all vented appliances (except for some types of incinerators) use some type of *draft hood* or *draft diverter*. Figure 4-2 at the top of the next page shows four common draft hood designs. Note that these draft hoods all have an inlet opening, which is connected to the flue outlet, and an outlet opening connected to the vent pipe. The common feature of a draft hood by which it differs from other vent fittings is an atmospheric opening exposed to the atmosphere. This atmospheric opening usually is located somewhere above the burner level.

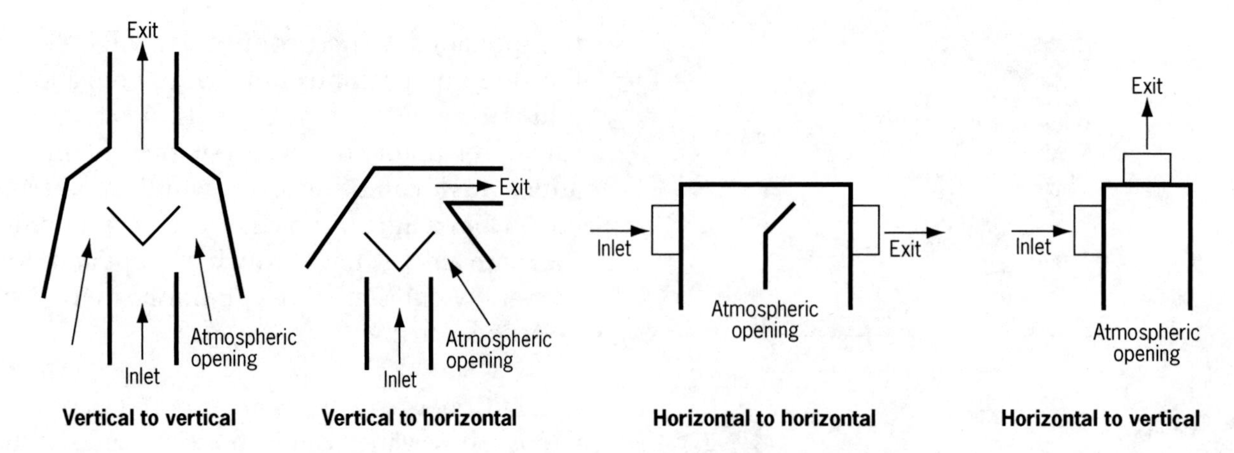

Figure 4-2. Typical gas appliance atmospheric draft hoods

The height of the draft diverter over the fire determines the "over-the-fire" draft. It is important with boilers not to modify the draft diverter height by cutting the vent connector that is connected to the draft diverter. Make sure that you get the correct draft diverter for the boiler and install it without modification.

Figure 4-3 below shows the operation of a typical draft hood under three conditions—normal (no wind), updraft, and downdraft. The circled numbers in Figure 4-3 represent the steps of the operation. The flow of (1) combustion products, (2) dilution air entering through the atmospheric opening, and (3) vent gases is indicated by the lengths of the arrows.

With normal venting, some air is pulled into the draft hood by the natural draft action of the gravity vent. This air mixes with the flue products from the furnace or boiler. Mixtures of combustion products, excess air, and dilution air drawn into the vent through the draft hood atmospheric opening are referred to collectively as *vent gases*. In a sense, the atmospheric opening "breaks" the gravity vent system at the draft hood. By doing so, it minimizes changes in stack action during normal burner operation.

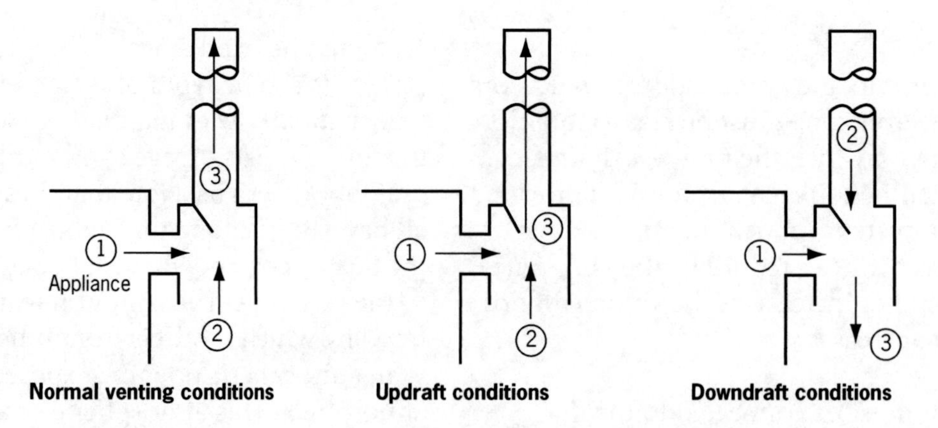

Figure 4-3. Operation of draft hood under various wind conditions

Updraft in a vent may be increased, either by using a higher stack, or intermittently by the action of wind. When updraft is increased, more cool air is drawn through the atmospheric opening into the stack. This action minimizes changes in air flow through the combustion area in two ways. First, it lowers the temperature of the vent gases, reducing gravity stack action. Second, any increase in air flow through the gravity vent because of increased updraft is made up mostly of dilution air (drawn into the stack through the atmospheric opening, rather than through the combustion area).

Under downdraft conditions, a draft hood prevents the buildup of positive pressure at the flue outlet. Ideally, a draft hood maintains atmospheric pressure at the flue outlet under downdraft conditions, and the gas burner operates as though there were no downdraft. With a draft hood in place, the burner can operate under temporary downdraft conditions by venting out through the draft hood atmospheric opening. The burner continues to be supplied with sufficient air in spite of the downdraft. When downdraft conditions cease, normal venting conditions resume. Normally, downdraft conditions are temporary and of short duration.

Fan-assisted systems include both forced-draft and induced-draft designs. A *forced-draft* system places a fan at the inlet side of the burner to push air *into* the combustion chamber, creating a positive pressure. An *induced-draft* system places a fan at the outlet of the combustion chamber to pull combustion gases *out*, creating a negative pressure in the combustion chamber. Fan-assisted systems allow for precise control of the amount of combustion air by eliminating outside influences that cause the amount of excess air in the flue to fluctuate. This more precise control improves the efficiency and is used in most of today's new furnaces.

VENTING CATEGORY I APPLIANCES

The purpose of a vent system is to completely remove all gases to the outside without allowing condensation in the vent or spillage at the draft hood. Technicians involved in installing a vent connector for gas appliances should be familiar with and understand the venting guidelines for Category I appliances as set forth in NFPA 54/58, and follow the manufacturer's installation instructions. Before making flue connections, consult local gas company requirements and any local codes that may apply. Then proceed with the following recommendations:

1. The vent connector must be the same diameter as the flue connection of the unit.
2. The vent connector shall run as directly as possible, with a minimum number of turns.
3. The pipe must have a minimum upward slope of at least 1/4 in. per linear foot for horizontal runs. If several appliances are connected together, then the vent connector that joins the appliances shall be installed in accordance with the manufacturer's instructions and the venting requirements of NFPA 54/58.
4. The vent connector shall extend through the chimney wall, but not beyond the inside wall of the flue.
5. The vent connector must be adequately supported. It shall not run to a chimney that serves an open fireplace.
6. When extending through the roof, the vent connector shall be equipped with a weather hood.
7. Chimneys or vents that run through the roof should extend at least 3 ft above the roof penetration and at least 2 ft above any surface within a 10-ft horizontal plane, or as required by code for pitched roofing.
8. Venting systems that run through walls and roof assemblies must meet requirements for structural combustion materials.

9. Where two or more appliances vent into a common flue, the area of the common flue shall be sized according to NFPA 54/58 and applicable manufacturer installation instructions.

FLUE SYSTEM DESIGN

Proper installation and maintenance are both important for the correct operation of a flue system. The design and installation should include the following factors:

- To prevent downdrafts caused by the wind moving across the roof, the top of a chimney must extend at least 2 ft above the highest point of the roof. The chimney top must be at least 3 ft higher than the point at which it passes through the roof (see Figure 4-4).
- Where it enters the chimney, the vent connector must not extend beyond the inner face or the liner of the chimney (see Figure 4-5).
- Horizontal vent connector runs should be kept as short as possible.
- The vent connector must be supported and joints must be tight and secure. Joints should be made in the direction of flow to reduce resistance (see Figure 4-6).
- Where local experience indicates, provisions must be made to prevent condensate from running into the vent connector and corroding it (see Figure 4-7). This arrangement applies not only to Categories II and IV, but also to Categories I and III.
- Both the chimney and vent connectors must be clear of all obstructions. Sources of blockage, as shown in Figure 4-8, may include:
    - a damaged rain cap on the chimney
    - a bird's nest
    - debris that may have entered the chimney
    - displaced brick in a masonry chimney
    - a loose or open joint
    - dented, damaged, or displaced vent connectors.

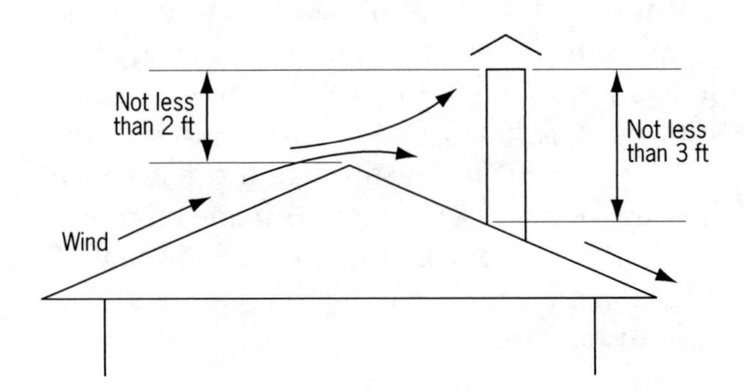

Figure 4-4. Chimney requirements

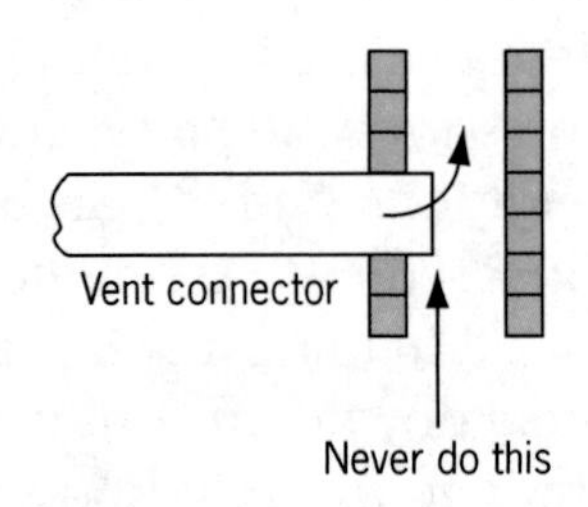

Figure 4-5. Vent connector extension

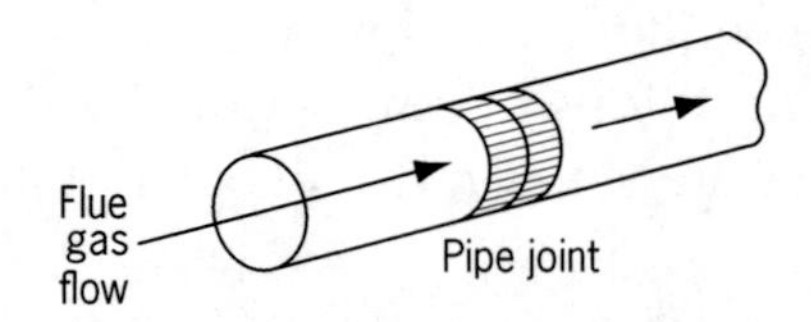

Figure 4-6. Secure joints

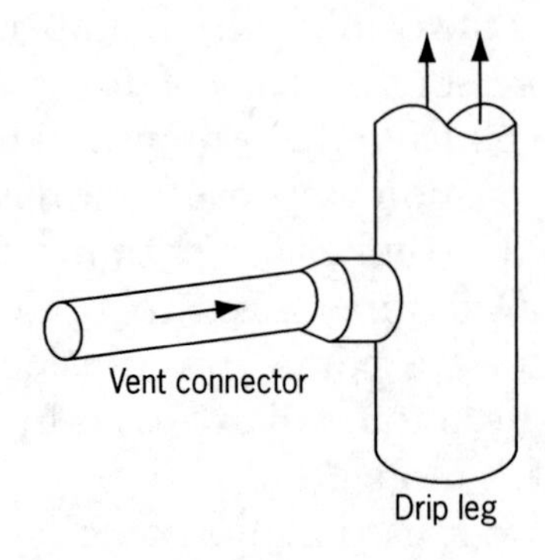

Figure 4-7. Vertical drip leg

- Horizontally run piping must have an upward pitch of at least 1/4 in. per foot of length.
- Occasionally other burners that are vented into the same vent connector or chimney will overload it and rob it of its capacity to serve the furnace. If this is suspected, check by turning off the other burners and temporarily blocking their connections into the flue system.
- Poor draft conditions, excessive flue height, or excessive horizontal runs can cause condensation to occur in the vent connector. To reduce the potential occurrence of condensation, vent connectors shall be of Type B, Type L, or listed vent material having equivalent insulation qualities. A "Type B" vent is a double-wall metal vent pipe that provides insulation to help prevent condensation. It is important to note that not all manufacturers use the same locking joint. If part of the pipe must be replaced or extended, a compatible brand must be used.
- A vent connector should have the least possible number of elbows and turns. Using too many of them creates friction losses.

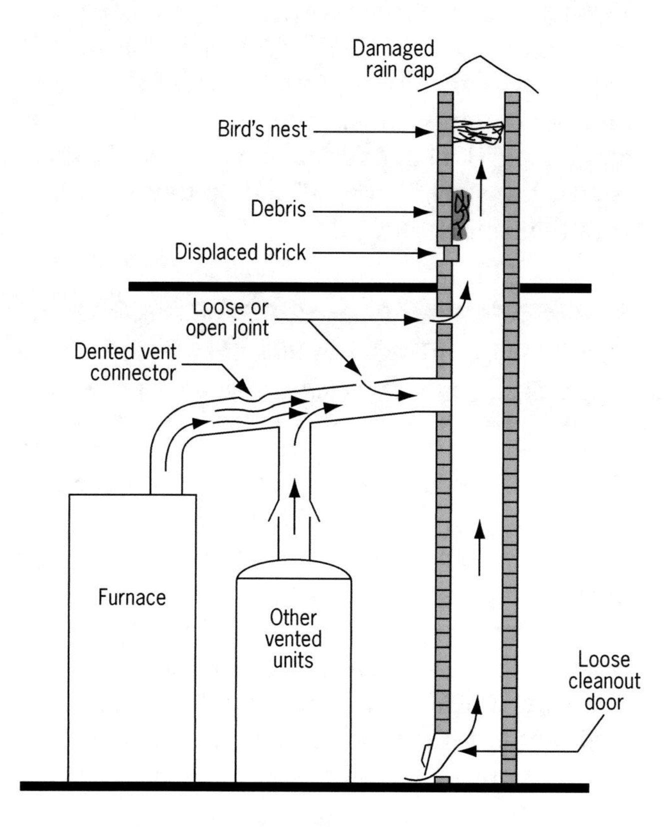

Figure 4-8. Chimney and vent connector obstructions

Energy-efficient houses that are well-insulated and tightly constructed can cause problems with venting. A "tight" house may not have enough air infiltration for combustion, dilution, and ventilation air. Other appliances—such as water heaters, clothes dryers, and exhaust fans—also can rob the furnace of combustion air. As a result, it may be necessary to install a combustion air pipe of sufficient size to supply proper combustion air for the furnace. The combustion air pipe should terminate close to the fossil fuel furnace and other fossil fuel appliances.

If incomplete combustion due to a lack of combustion air is suspected, the system must be run at steady state for a period of time. Measurements for carbon monoxide must be taken before the draft diverter (if there is one). If there is a draft inducer, the carbon monoxide measurement must be taken at the discharge of the draft inducer.

VENTING CATEGORY IV (CONDENSING) EQUIPMENT

Direct venting

A *direct* venting system is one that draws 100% of the combustion air directly from the outside through a sealed pipe that runs from the outside to the combustion chamber. This type of system prevents the possibility of the burner being starved by exhaust fans, tight construction, and other fuel-burning appliances within the same space.

Direct venting eliminates contaminants (such as chemicals or lint from a dryer) from the indoor space. It also eliminates the need for a combustion air duct into the space. This arrangement requires two pipes (see Figure 4-9), but it is the best way to ensure that a sufficient amount of clean combustion air is delivered to the burner. *Note:* Piping must be installed according to the equipment manufacturer's requirements.

Non-direct venting

A *non-direct* venting system draws combustion air from the space in which it is installed. It is very important to consider the location and construction of the space carefully to make sure that sufficient combustion air is able to enter the space.

If it is possible for the building to be placed in a negative pressure due to the operation of exhaust fans or other combustion equipment, this method should not be used. If the environment contains significant amounts of dust, lint, or chemicals that could be drawn into the combustion air, direct venting should be used.

CHIMNEYS

Generally speaking, an existing chimney cannot be used to vent Category IV appliances. A typical chimney has no provision for dealing with the condensate, which means that it could drip from the chimney into the occupied space. The condensate can be very corrosive, and can quickly destroy the masonry and steel materials typically used in chimneys. In addition, code regulations do not allow the common venting of Category IV appliances along with any Category I appliances that may be using the chimney. An abandoned chimney may be used as a chase in which to run new vent lining, as long as the new vent lining runs the entire length of the original chase.

Figure 4-9. Direct venting

MATERIALS

It is important that the manufacturer's instructions be followed in determining the proper materials for the venting system of an appliance. One of the most critical considerations in selecting a material is the corrosive nature of the condensate. Most applications use PVC pipe for both combustion air intake and venting. The manufacturer also may use other materials, such as stainless steel, that will hold up to the contact with the condensate.

National Fuel Gas Code 12.5.2
"Plastic piping used for venting appliances listed for use with such venting materials shall be approved."

Category IV appliances use a fan to force the combustion gases through the vent piping. Because the fan creates a positive pressure within the vent pipe, any unsealed joints will allow flue gases to leak outward. Technicians must make certain that all joints are properly sealed and that there are no leaks. PVC pipe should be glued at all joints. Other materials typically provide a gasketed or welded joint to ensure a tight seal.

Some special fittings may be available to make the installation of the venting system easier or to

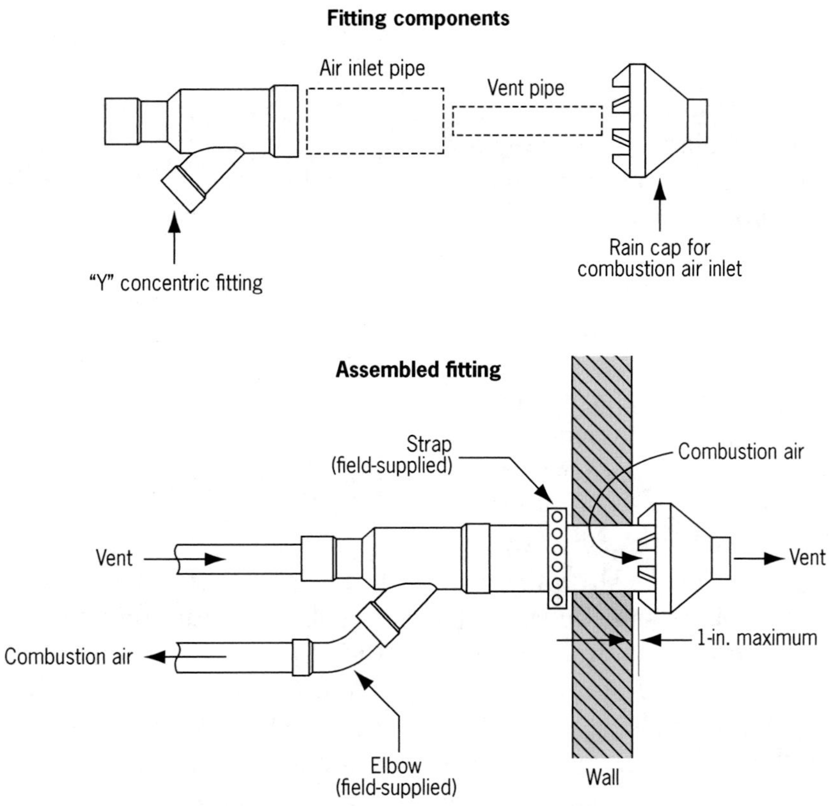

Figure 4-10. Concentric fitting

improve its appearance. One such fitting that is frequently used is a *concentric* fitting. As you can see in Figure 4-10, it allows the vent pipe and the air intake to exit the building through one penetration. The fitting is actually a double-wall pipe with the flue gases in the middle and the air intake around the outside.

PITCH

Pitch is very important to vent piping used with Category IV equipment. Since the normal operation of the equipment causes condensate to form in the piping, some method must be provided to remove that condensate. A typical installation maintains an upward pitch from the equipment to the termination outside the building. As condensate forms in the vent pipe, it drains back to the equipment and out to a proper drain. A low spot in a vent line forms a trap where the condensate collects and prevents flue gases from passing through.

Another important reason for pitching the pipe properly is to prevent large amounts of ice from forming on the end of the discharge pipe, which also can restrict the flow of flue gases. Always consult the manufacturer's instructions for the amount of pitch required for a given installation.

National Fuel Gas Code 12.10.1

"Provision shall be made to collect and dispose of condensate from venting systems serving Category II and Category IV appliances and noncategorized condensing in accordance with NFPA 12.9.4."

SIZING

Proper sizing of the vent piping is a very important part of the installation. An *undersized* vent connector creates too much restriction and prevents the proper amount of air from being moved by the induced-draft fan. An *oversized* pipe slows the velocity of the flue gases and causes a greater volume of condensate to form.

All major equipment manufacturers provide sizing tables. Table 4-1 on the next page shows an extract from one such manufacturer-specific table. You must determine the total length of pipe and the number of 90° elbows that will be needed before using the table. The hypothetical system shown below in Figure 4-11, for example, requires 28 ft of pipe and four elbows. You also know that the furnace in this example has an input rating of 80,000 Btuh. To use the table, match this figure to the equipment rating in the second column. (Note that the first column lists altitude. High-altitude applications require some adjustments.)

When you have found the horizontal row with the 80,000-Btuh input rating (see shaded box), follow it to the far right-hand columns, which list the number of elbows (1 through 6). In the "4" column (because there are four elbows in our example), you will see the maximum length of pipe that can be used. Find the figure that is closest to and greater than the needed length. Our example calls for 28 ft of pipe, so the next higher number is 30 ft (see shaded box). Now follow that row back to the left to find the proper pipe size (diameter) for the application, which is 2 in. in this case (see shaded box). If the length of the vent piping exceeds that which is listed in the table, you will need to contact the manufacturer.

INSULATION OF THE VENT PIPING

If vent piping is run through an unconditioned space, such as a crawl space or attic, there is a possibility of the condensate freezing inside the vent and restricting the flow of flue gases. Table 4-2 on page 48 lists the maximum length of pipe that can be exposed to an unconditioned space for various sizes of equipment, winter design temperatures, and pipe diameters. If the length of pipe that must run through the

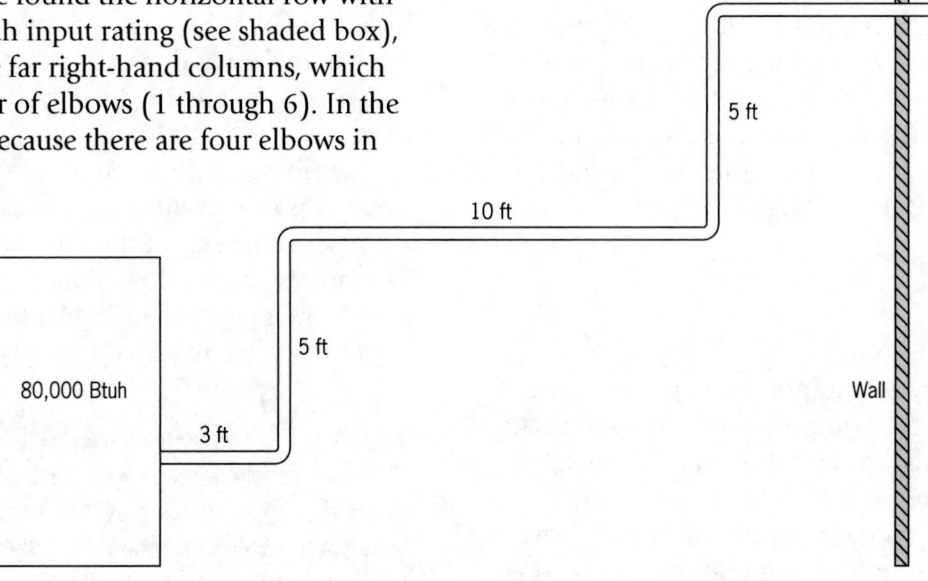

Figure 4-11. Vent piping sizing example

NOTE: TABLE APPLIES TO CARRIER APPLIANCES ONLY.

Altitude (ft)	Unit maximum input rate (Btuh)	Direct vent (2-pipe only): Termination type	Direct vent (2-pipe only): Pipe diameter (in.)	Non-direct vent (1-pipe only): Pipe diameter (in.)	Maximum allowable pipe length (ft) — Number of 90° elbows: 1	2	3	4	5	6
0–2,000	60,000	2-pipe or 2-in. concentric	1½	1½	20	15	10	5	NA	NA
			2	2	70	70	70	70	70	70
	80,000	2-pipe or 2-in. concentric	1½	1½	10	NA	NA	NA	NA	NA
			2	2	55	50	35	30	30	20
			2½	2½	70	70	70	70	70	70
	100,000	2-pipe or 3-in. concentric	2	2	5	NA	NA	NA	NA	NA
			2½	2½	40	30	20	20	10	NA
	120,000	2-pipe or 3-in. concentric	2½ one disk	2½	10	NA	NA	NA	NA	NA
			3*	NA	45	40	35	30	25	20
			3* no disk	3*	70	70	70	70	70	70
2,001–3,000	60,000	2-pipe or 2-in. concentric	1½	1½	17	12	7	NA	NA	NA
			2	2	70	67	66	61	61	61
	80,000	2-pipe or 2-in. concentric	2	2	49	44	30	25	25	15
			2½	2½	70	70	70	70	70	70
	100,000	2-pipe or 3-in. concentric	2½	2½	35	26	16	16	6	NA
			3	3	70	70	70	70	66	61
	120,000	2-pipe or 3-in. concentric	3	NA	14	9	NA	NA	NA	NA
			NA	3*	63	62	62	61	61	61
			3* no disk	NA	70	70	63	56	50	43
			4* no disk	4* no disk	70	70	70	70	70	70
3,001–4,000	60,000	2-pipe or 2-in. concentric	1½	1½	16	11	6	NA	NA	NA
			2	2	68	63	62	57	57	56
	80,000	2-pipe or 2-in. concentric	2	2	46	41	28	23	22	13
			2½	2½	70	70	70	70	70	70
	100,000	2-pipe or 3-in. concentric	2½	2½	33	24	15	14	5	NA
			3	3	70	70	70	66	61	56
	120,000	2-pipe or 3-in. concentric	3* no disk	NA	65	58	51	44	38	31
			NA	3*	59	59	58	57	57	56
		4* no disk	4* no disk	4* no disk	70	70	70	70	70	70

*See explanatory notes at end of manufacturer's table.

CARRIER CORPORATION

Table 4-1. Combustion air and vent piping for direct vent (2-pipe) and non-direct vent (1-pipe) applications (extract from Carrier 58MTB)

unconditioned space exceeds that listed in the table, insulation should be added to the pipe according to the manufacturer's requirements.

TERMINATION

The proper termination of the vent piping is important to consider before starting any venting job. Category IV vents can exit the building either vertically through the roof or horizontally through the sidewall. Terminations must follow the manufacturer's requirements for proper spacing. The termination must be away from any air opening into the building to prevent any of the combustion products from entering the building.

NOTE: TABLE APPLIES TO CARRIER APPLIANCES ONLY.

CARRIER CORPORATION

Unit maximum input rate (Btuh)	Winter design temperature* (°F)	Maximum pipe diameter (in.)	Without insulation	With 3/8-in. or thicker insulation*
60,000	20	2	44	70
	0	2	21	70
	-20	2	20	57
80,000	20	2	55	55
	0	2	30	55
	-20	2	16	55
	20	2½	58	70
	0	2½	29	70
	-20	2½	14	67
100,000	20	2½	40	40
	0	2½	38	40
	-20	2½	21	40
	20	3	63	70
	0	3	30	70
	-20	3	12	70
120,000	20	3	70	70
	0	3	38	70
	-20	3	19	70
	20	4	65	70
	0	4	26	70
	-20	4	5	65

*See explanatory notes at end of manufacturer's table.

Table 4-2. Maximum allowable exposed vent pipe length (ft) with and without insulation in unconditioned spaces (extract from Carrier 58MTB)

National Fuel Gas Code 12.9.3

"The vent terminal of a direct vent appliance with an input of 10,000 Btuh or less shall be located at least 6 in. from any air opening into a building, an appliance with an input over 10,000 Btuh but not over 50,000 Btuh shall be installed with a 9-in. vent termination clearance, and an appliance with an input over 50,000 Btuh shall have at least a 12-in. vent termination clearance. The bottom of the vent terminal and the air intake shall be located at least 12 in. above finished ground level."

National Fuel Gas Code 12.9.4

"Through-the-wall vents for Category II and Category IV appliances and noncategorized condensing appliances shall not terminate over public walkways or over an area where condensate or vapor could create a nuisance or hazard or could be detrimental to the operation of regulators, relief valves, or other equipment. Where local experience indicates that condensate is a problem with Category I and Category III appliances, this provision shall also apply. Drains for condensate shall be installed in accordance with the appliance and vent manufacturers' installation instructions."

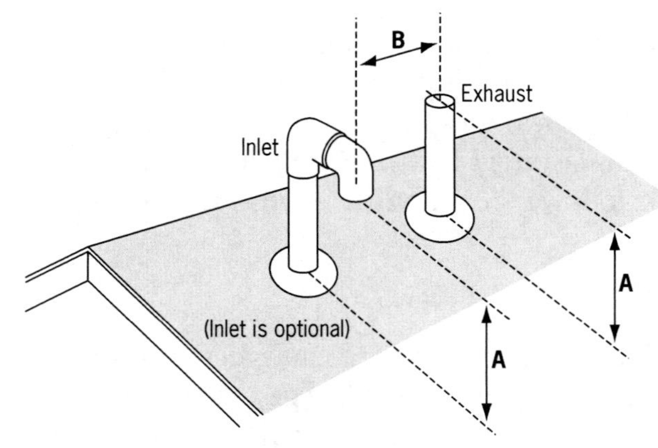

Rooftop termination

A = 12 in. above roof or snow accumulation level
B = 8-in. minimum, 20-in. maximum, except in areas with extreme cold temperatures (sustained below 0°F), then 18-in. minimum

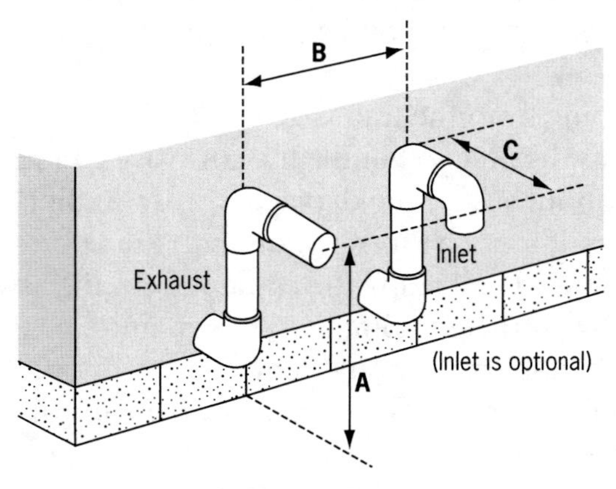

Sidewall termination

A = Minimum 12 in. above expected snow accumulation level
B = 8-in. minimum, 20-in. maximum, except in areas with extreme cold temperatures (sustained below 0°F), then 18-in. minimum
C = 8-in. minimum

Figure 4-12. Termination of vent piping

Terminations also must be located so that condensate that may drip from the end of the pipe will not create a problem. The vent may not terminate over a public walkway. In addition, the vent may not terminate over other equipment, such as a gas meter or gas regulator, where the corrosive nature of the condensate could potentially damage the equipment or the moisture could freeze on the equipment and prevent it from functioning properly (e.g., ice plugging a vent hole on a gas regulator or some type of relief valve).

Snow must be taken into consideration when placing the vent piping. All piping should terminate at least 12 in. above the expected snow level to prevent blockage due to snowfall (see Figure 4-12). Clearance to other vent terminations also must be observed. *Not all manufacturers have the same requirements, so it is important to check the instructions for the specific equipment.*

Direct vent applications may use concentric terminations. This type of termination uses a double-wall pipe, with flue gases carried in the inner pipe and combustion air carried in the outer pipe. Both pipes exit through one hole, providing a neat, clean termination. The manufacturer's installation instructions must be followed for all installations.

CONDENSATE DRAINS

Since Category IV equipment is designed to permit the water in the flue gases to condense, a drain must be provided to remove the condensate. The manufacturer of the equipment normally provides a trap to collect the condensate, as well as a seal between the venting system and the drain piping. Check the manufacturer's requirements for priming the condensate trap.

Piping from the trap is generally PVC because it must be able to handle the corrosive condensate without sustaining damage. The drain pipe should not be piped solidly into the sewer—doing so will allow the effluent from the sewer to enter the equipment in the event of a backup. The drain pipe should terminate at an open-site drain, such as a floor drain, with an air gap as shown in Figure 4-13.

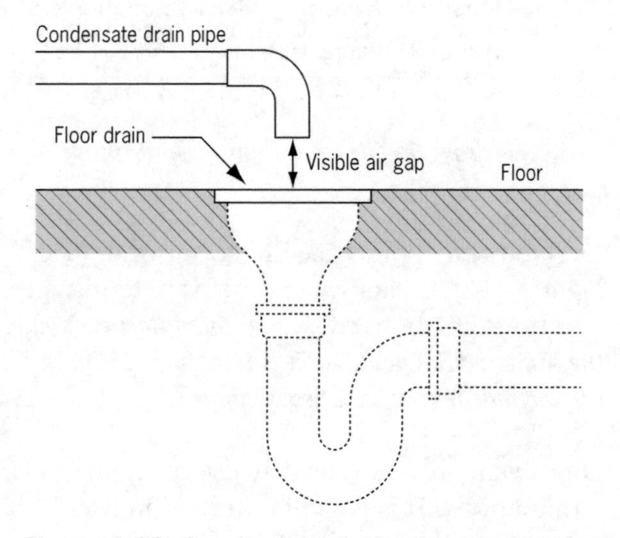

Figure 4-13. Condensate drain pipe termination

CODES

The United States and Canada follow very similar codes, which basically state that all venting must meet the manufacturer's requirements. It is very important to consult the local authority having jurisdiction (AHJ) and conform to all applicable codes for your specific area. *Note that this Chapter provides basic venting guidelines only. There is no substitute for reading the equipment manufacturer's installation instructions and following all of the listed requirements.* □

REVIEW QUESTIONS

1. A standing pilot furnace with an atmospheric burner is an example of a _________ gas appliance.

   a. Category I
   b. Category II
   c. Category III
   d. Category IV

2. A high-efficiency condensing furnace is an example of a _________ gas appliance.

   a. Category I
   b. Category II
   c. Category III
   d. Category IV

3. Burning of 1 ft^3 of natural gas produces _________ of flue gas.

   a. 1 ft^3
   b. 2 ft^3
   c. 10 ft^3
   d. 11 ft^3

4. Which of the following affects the venting force in a gravity venting system?

   a. The height of the chimney
   b. The percentage of CO_2 in the flue gas
   c. The type of fuel used
   d. The type of fuel pipe used

5. A fan is used to pull flue gases through the combustion chamber in a(n) _________ vent system.

   a. forced-draft
   b. gravity
   c. induced-draft
   d. natural draft

6. How far above the peak of the roof should the chimney extend?

   a. 1 ft
   b. 2 ft
   c. 3 ft
   d. 4 ft

7. A horizontal flue vent must have an upward slope of at least _________ per foot of length.

   a. 1/8 in.
   b. 1/4 in.
   c. 1/2 in.
   d. 1 in.

8. Problems related to a shortage or contamination of combustion air are eliminated in a(n) _________ vent system.

   a. direct
   b. forced-draft
   c. gravity
   d. induced-draft

9. When can a masonry chimney be used to vent Category IV appliances?

   a. When it is an existing chimney
   b. When it is no more than one story tall
   c. When it is used as a common vent along with Category I appliances
   d. Never

10. Category IV furnaces are commonly vented with what type of material?

    a. Single-wall metal
    b. Double-wall Class B
    c. Masonry
    d. PVC pipe

11. When does a Category IV vent pipe require insulation?

    a. Always
    b. Only when the vent pipe runs outdoors
    c. When the length of the vent pipe running through an unconditioned space exceeds the length listed in the manufacturer's tables
    d. Never

12. Category IV equipment requires a condensate drain. A trap must be installed in the drain line to _________.

    a. break the vacuum and allow the water to drain
    b. keep the water from draining too quickly
    c. prevent the water from freezing in the vent connector
    d. provide a seal that prohibits the leakage of flue gas

ILLUSTRATION CREDITS
FIGURE 4-9: RICH HOKE
FIGURE 4-10: CARRIER CORPORATION

CHAPTER 5

Air Flow

INTRODUCTION

In order for a furnace to complete the job of heating a space, the proper amount of air must move through the furnace to pick up the heat from the heat exchanger and deliver it to the conditioned space. Too little air moving through the system will cause the equipment to overheat, tripping the limit safety and shutting down the furnace. Frequent cycling on and off ultimately shortens the life of the equipment. Insufficient air flow also can cause distribution problems and uneven temperatures within the space. Too much air flow, on the other hand, increases the amount of energy being consumed by the blower motor and often results in a drafty and noisy distribution system.

MEASURING AIR FLOW

Before measuring air flow, remember to check the blower. If the blades on the blower wheel are bent or packed with dirt, the amount of air that it can deliver will be greatly reduced. The air filter also should be checked and replaced as needed.

The following paragraphs describe three methods that can be used to measure air flow and ensure

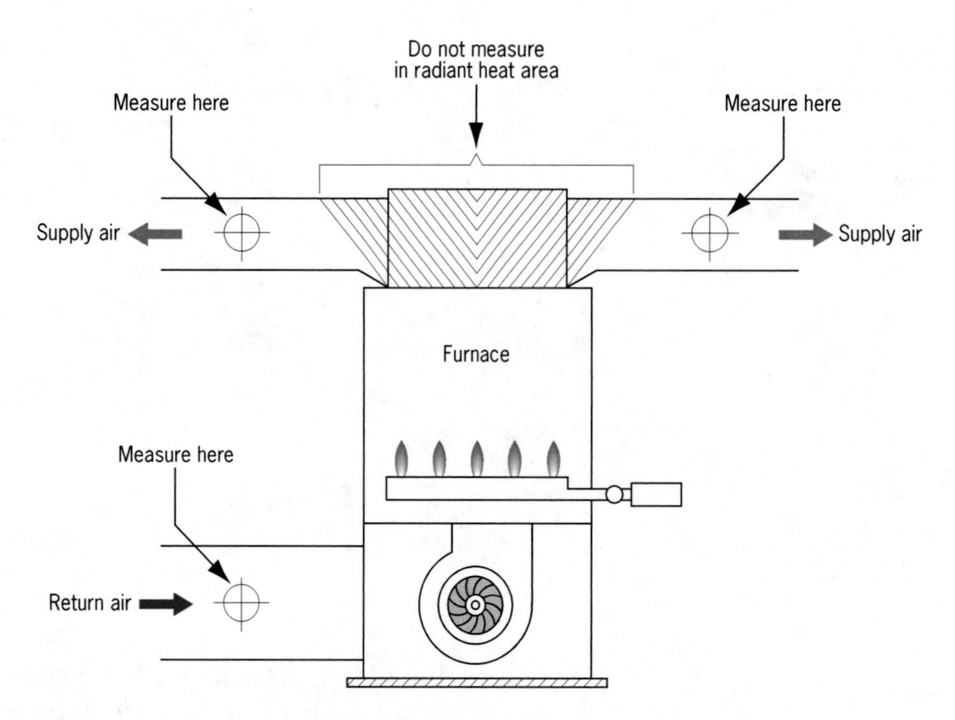

Figure 5-1. Measuring temperature rise

that the furnace is operating within the proper range:

- temperature rise
- blower performance data
- direct measurement.

Each of these methods has its advantages and limitations.

Temperature rise

The temperature rise method is commonly used with heating equipment. Before measuring temperature rise, you must first make sure that the burner output is correct. Start by checking the fuel pressure to the burner. Once you have confirmed that the manifold pressure is correct, clock the meter (either dial or digital) to verify the proper Btu rating. (This applies only to natural gas furnaces—the temperature rise method *cannot* be used with air conditioning or heat pump units, since the output of the evaporator coil is a combination of both sensible and latent cooling in the cooling mode, and in the heating mode the capacity of the heat pump varies with the outside temperature.)

An electric heating element works best for temperature rise measurements, because the Btu output can be easily calculated. In order to determine power, simply measure current and voltage (amperes × volts = watts), and then multiply by 3.41 to convert watts to Btus.

Before attempting to use the temperature rise method, first make sure that the furnace has been running long enough to achieve steady-state conditions. Then measure the supply air temperature in the supply duct. Take your reading as close as possible to the heat exchanger without being in its "line of sight." This prevents radiant heat from the heat exchanger from influencing the temperature reading (see Figure 5-1 above). Next, measure the return air temperature in the return duct connection to the furnace. It is a good idea to use the same thermometer to take both measurements. (This way, if the thermometer that you are using *is* out of calibration, both readings will be off by the same amount, and the temperature difference will still be correct.) The difference in temperature between the two points is called the *temperature rise*.

Air flow, cfm	External static pressure (in. w.c.)											
	Vertical*						Horizontal**					
	230 V			208 V			230 V			208 V		
	HI	MED	LO	HI	MED	LO	HI	MED	LO	HI	MED	LO
500						0.55						
550						0.51						0.60
600					0.67	0.41						0.58
650			0.54		0.60	0.23			0.60			0.51
700			0.53		0.52	0.00			0.57		0.51	0.47
750		0.48	0.44	0.65	0.41			0.54	0.53		0.48	0.35
800	0.52	0.47	0.27	0.59	0.30		0.60	0.52	0.46	0.59	0.41	0.05
850	0.50	0.41	0.00	0.52	0.10		0.57	0.47	0.32	0.55	0.32	
900	0.47	0.30		0.42	0.01		0.54	0.40	0.03	0.52	0.21	
950	0.41	0.15		0.29			0.49	0.31		0.45	02	
1000	0.33	0.00		0.14			0.41	0.19		0.33		
1050	0.22			0.00			0.32	0.04		0.19		
1100	0.10						0.23			0.00		
1150	0.00						0.12					
1200							0.02					

* Vertical installation: With filter, no horizontal drip tray. Small apex baffle. Subtract 0.06 in. w.c. for downflow.
** Horizontal installation: As shipped, but without filter. Subtract 0.05 in. w.c. for horizontal left.

TRANE/AMERICAN STANDARD INC.

Table 5-1. Blower performance data for typical 2½-ton indoor unit

By checking your result against the equipment nameplate, you can determine if the air flow is within the correct range. If the measured temperature rise is higher than the nameplate rating, there is not enough air moving through the heat exchanger. If the measured temperature rise is lower than the nameplate rating, there is excess air flow.

To calculate the actual air flow across the furnace, you can use the following equation:

$$\text{cfm} = \frac{\text{heating capacity (Btuh)}}{1.08 \times \text{temperature rise}}$$

For the heating capacity of the furnace, be sure to use the Btuh *output* rating, which is usually marked on the unit's nameplate.

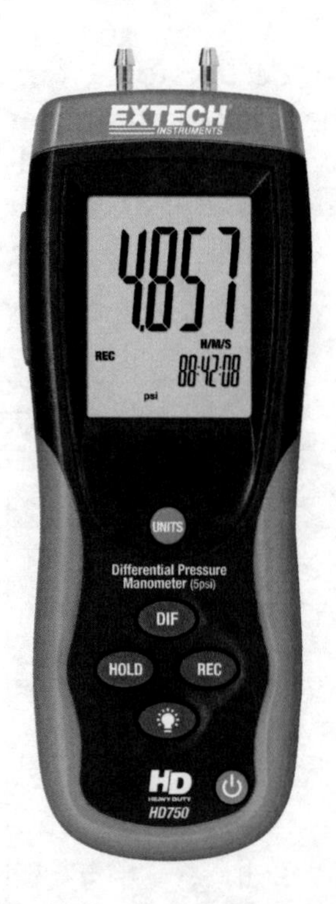

Figure 5-2. Digital manometer

Blower performance data

Blower curves and related data are often printed in tables in the manufacturer's service literature for smaller residential and light commercial equipment. Table 5-1 on the previous page shows an example of a published air flow performance table. It lists static pressures at various fan speeds and tells you what the air flow is for the given conditions. Then you can compare the cfm number provided by the table to the actual equipment requirements. This method is especially useful for checking air flow in air conditioning and heat pump systems.

To use these tables, you first must measure the external static pressure on the unit. The *external static pressure* (ESP) is the resistance to air flow caused by the ductwork, dampers, grills, filters, coils, and any other devices located in the airstream. Most residential units are designed to deliver the nominal cfm for an AHRI-rated match at an available ESP of up to about 0.5 in. w.c. If the ESP rises above this point, the blower may be able to deliver the needed cfm but it will require a higher blower speed, and thus more energy. Properly sized ductwork should be designed with friction rates of no more than 0.1 (in. w.c. static pressure drop per 100 ft of duct) on the supply side and 0.08 on the return side—although such "rule of thumb" values are not meant as a substitute for a proper *Manual D* friction rate calculation.

Any increase in the static pressure on either side of the system will cause a decrease in the total air flow. Dirty filters, obstructed coils, closed dampers, and blocked returns can increase static pressure—which, in turn, causes a reduction in air flow. Increasing duct sizes, adding return grilles, opening dampers, and providing clean filters will reduce static pressure and allow for an increase in air flow.

You can determine the total ESP by measuring the supply static pressure and return static pressure individually and adding the two values together—or, if your pressure gauge has two probes, simply by reading the pressure difference directly on the gauge. A handheld digital manometer similar to the one shown in Figure 5-2 or a diaphragm-type differential pressure gauge like the *Magnehelic® gauge* pictured in Figure 5-3 can be used for this task.

Figure 5-3. Magnehelic® gauge

In the example illustrated in Figure 5-4 below, the supply static pressure (SSP) is equal to 0.4 in. w.c. above atmospheric pressure and the return static pressure (RSP) is equal to 0.1 in. w.c. below atmospheric pressure. Note that the plus and minus signs (+ and –) are ignored and the values of the two pressures are simply added together to get the difference between the two readings. The ESP in this example, then, is equal to 0.5 in. w.c. (0.4 + 0.1).

Direct measurement

Measuring the air flow directly usually involves measuring the velocity of the air in the duct in feet per minute (ft/min) and multiplying that figure times the area of the duct in square feet (ft^2) to give you cubic feet per minute (cfm). An anemometer like the one shown in Figure 5-5 on the next page can measure air velocity directly in feet per minute.

It is also possible to determine air flow by measuring velocity pressure. *Velocity pressure* is the difference between the static pressure and the total pressure in the duct ($P_v = P_t - P_s$).

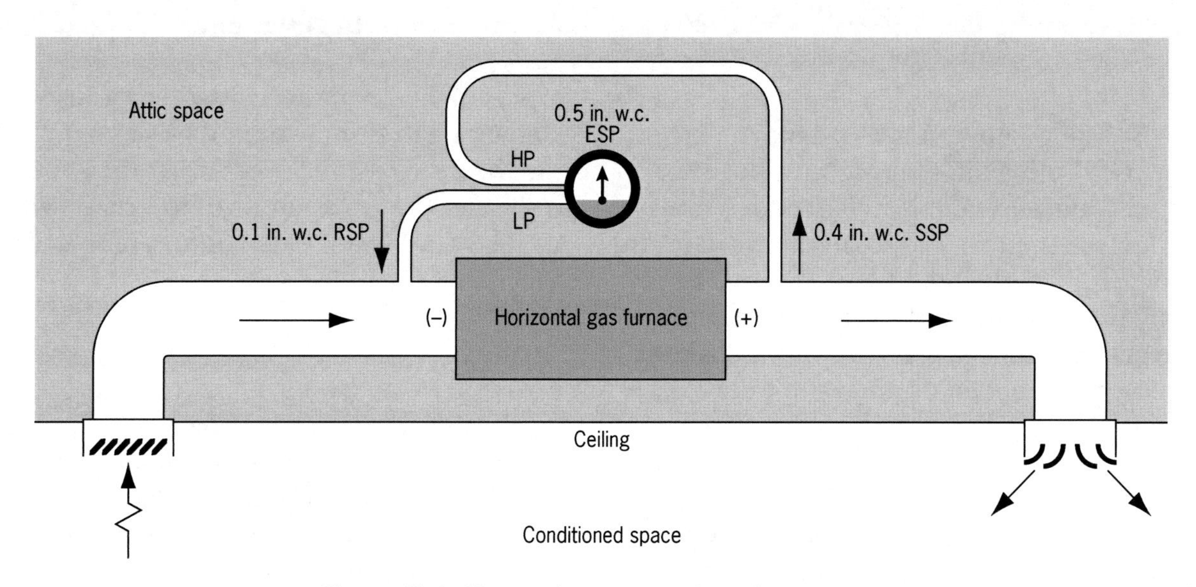

Figure 5-4. Measuring external static pressure

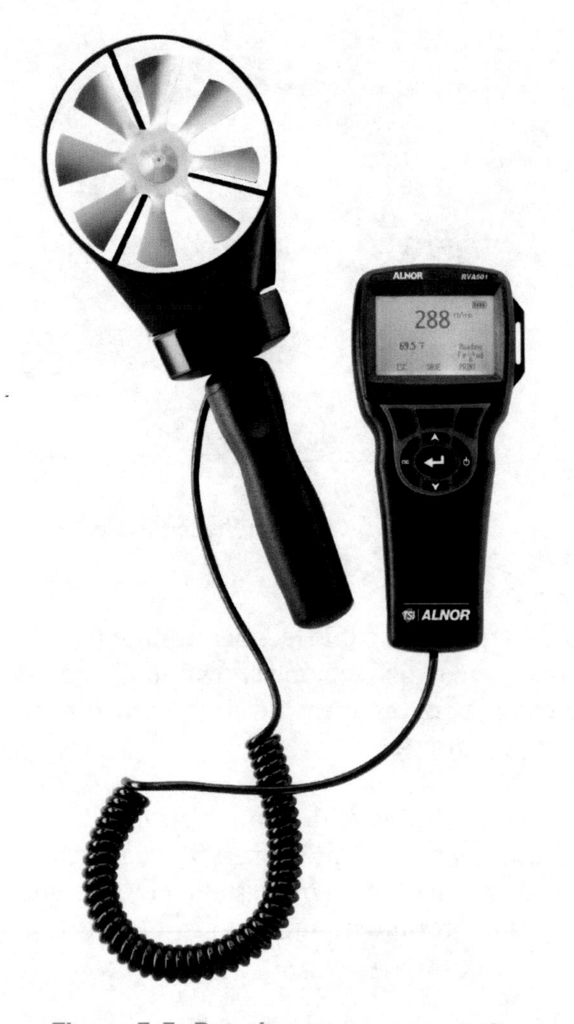

Figure 5-5. Rotating-vane anemometer

Figure 5-6 shows how an inclined manometer can be used with a Pitot tube to measure velocity pressure. Multiple readings, referred to as *traverse* readings (see Figure 5-7), usually are taken for the sake of accuracy. The results are converted from velocity pressure (in. w.c.) to velocity (ft/min) and averaged. That figure is then multiplied by the duct area to obtain the final cfm value.

RECOMMENDED AIR VELOCITIES

Recommended air velocities (ft/min) for various duct sections and applications are shown below. "Quiet areas" are broadcast studios, recording studios, and the like. "Face velocity" refers to the speed of the air as it emerges from a register or diffuser and enters the room. Increased velocity can cause excessive noise in the ductwork.

	Offices	Residences	Quiet areas
Trunk ducts	1200	1000	800
Branch ducts	800	600	500
Supply outlet face velocity	700	600	400

Recommended air speeds across filters are:

- disposable filters: 700 to 750 ft/min
- HEPA filters: 250 ft/min
- electronic air cleaners: 500 ft/min

Speeds higher than those listed will decrease filter efficiency.

Recommended air speeds across coils are:

- evaporators: 400 to 600 ft/min
- condensers: 1,000 ft/min
- hot water coils: 700 ft/min

Speeds higher than those listed may cause condensate to be blown off the evaporator coil into the duct system. Speeds lower than 400 ft/min may cause the evaporator coil to freeze. Refer to ANSI/ACCA *Manual D* to confirm proper air duct velocities.

DUCT SYSTEMS

The duct system is a critical part of delivering proper air flow, both to the equipment and to the space being conditioned. A duct system that is too small for the required air flow causes static pressure to increase—which in turn leads to a reduction in air flow and a noisy duct system.

In houses with forced-air heating and cooling systems, ductwork is used to distribute conditioned air throughout the house. In a typical house, however, about 20% of the air that moves through the duct system is lost due to leaks, holes, and poorly connected ducts. The result is higher utility bills and difficulty keeping the house comfortable, no matter how the thermostat is set.

Ductwork *must* be properly sealed and supported. Adequate support is necessary because sagging can open joints. Using the proper joining methods is also critical to making sure that joints remain as leak-free as possible. Figures 5-8 and 5-9 on the next page illustrate common joining methods for sheet metal and ductboard.

The correct type and placement of diffusers is important for proper air distribution. Since warm air rises naturally due to convection, the air temperature in a room can vary several degrees between top and bottom without good distribution. As a consequence, the air is significantly warmer near the ceiling than it is near the floor.

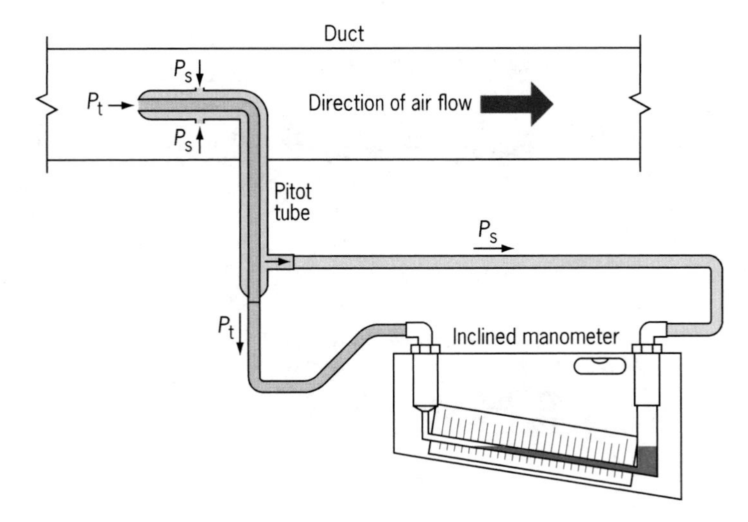

Figure 5-6. Pitot tube connected to manometer for measuring velocity pressure

In areas where the need for heat is predominant, diffusers generally should be placed in the floor. This arrangement takes advantage of natural convection to help circulate the air within the room. The opposite is true in a system that is used mostly for cooling—the diffusers are placed high

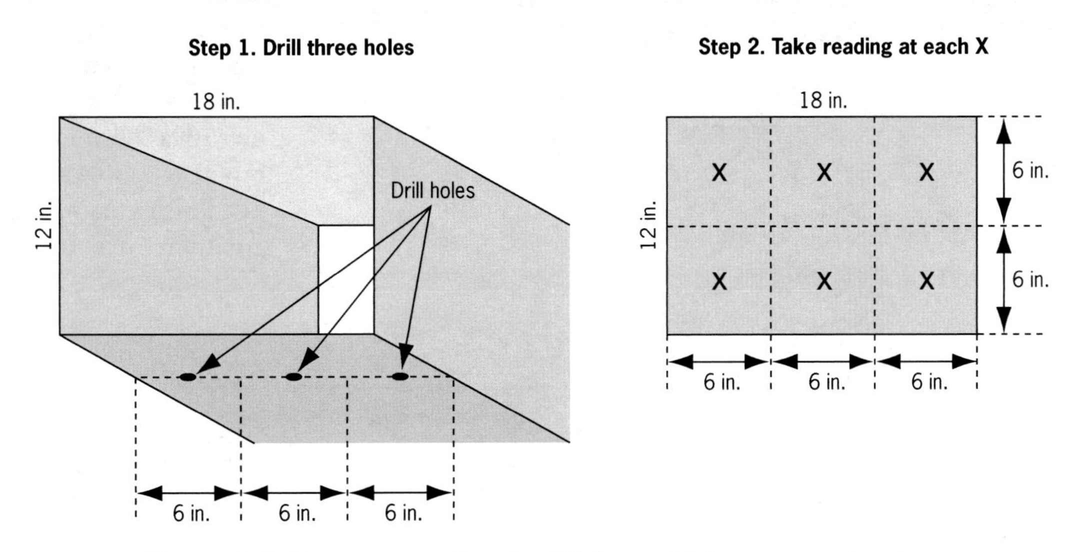

Figure 5-7. Traverse reading for 12-in. × 18-in. rectangular duct

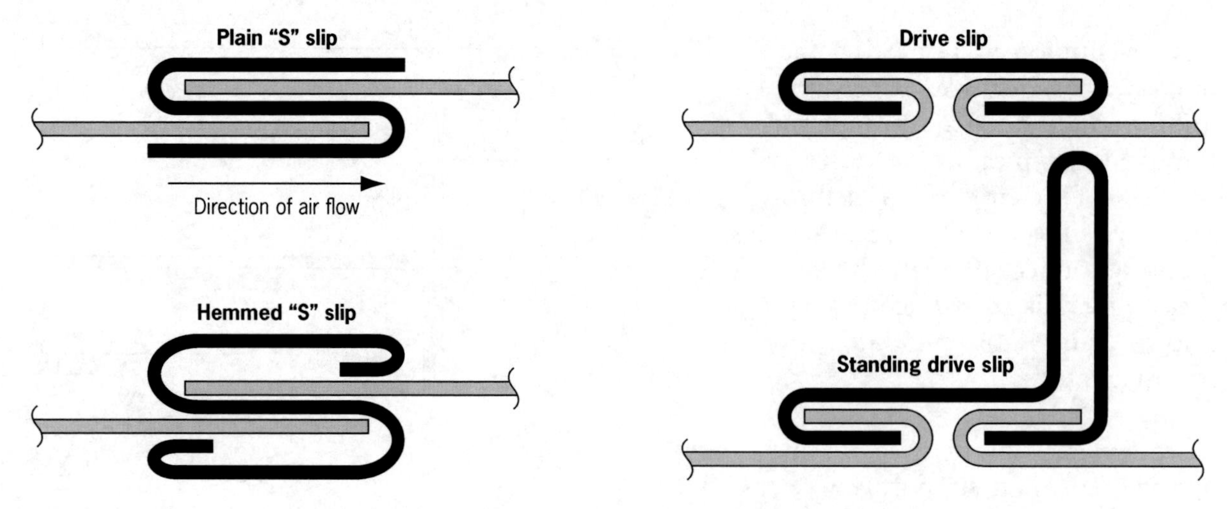

Figure 5-8. Common sheet metal duct joint designs

in the wall so that the cooler air will circulate downward and provide for natural circulation.

A diffuser's face velocity also plays a part in room air distribution. Insufficient velocity prevents the diffuser from achieving the proper *throw*, and the result is poorly mixed room air. If the velocity at the diffuser is too high, air delivery will be drafty and noisy. Diffusers also can get noisy and rattle if the built-in dampers become loose. They sometimes need to be replaced in sensitive areas.

Insulation is an important consideration in a duct system. Significant amounts of heat can be lost when ductwork is run through an unconditioned space. Ducts should be covered with insulation that is appropriate for the geographical location. Insulation placed on the inside of a duct not only provides thermal insulation, but also acts as a sound-deadening material and can greatly reduce noise in the ductwork.

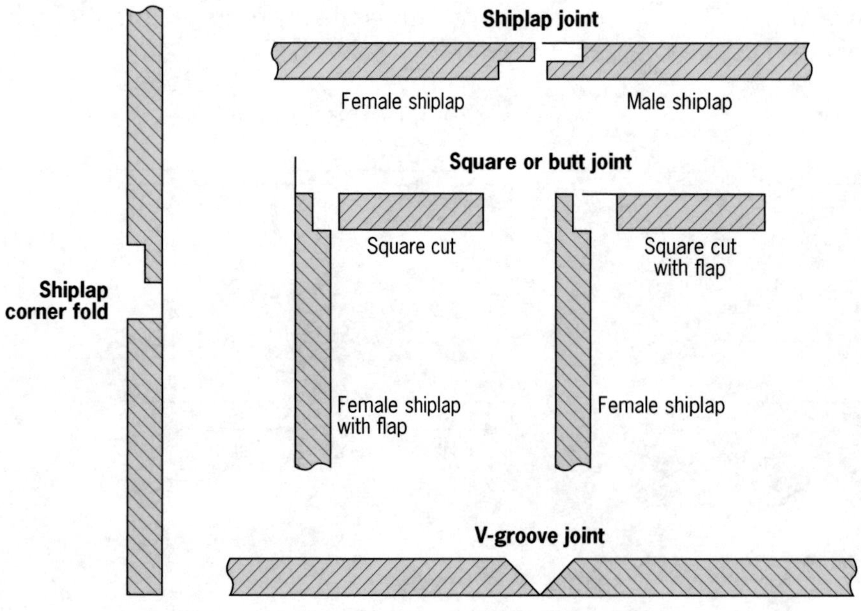

Figure 5-9. Ductboard joints

FILTERS

Filters are another important component of an efficient air distribution system. Historically, the filter's first job was to protect the *equipment* from large amounts of dirt buildup. Just as critical today is the role that filters play in protecting *people*—by reducing the number of contaminants in the air we breathe. As shown in Figure 5-10, filters are available in a large variety of types, depending on the intended application. Most filters are designed to remove particulates from the air. The removal of vapors or odors requires special adsorption-type filters or source control measures.

Filters have variable levels of filtration capability and resistance to air flow. Today's filters are rated with a standard testing method known as MERV (*minimum efficiency reporting value*). The higher the MERV rating, the better the filter is at catching small particles. The MERV rating provides an easy way to compare the many different types of filters. For residential applications, ANSI/ASHRAE Standard 62.2-2007 requires a filter with a designated minimum efficiency of MERV 6 or better.

Filter size should be selected based on total cfm and pressure drop requirements. Note that a filter with a higher MERV rating generally has a significantly higher pressure drop than a filter of the same size with a lower MERV rating. If you want to provide better filtration for a system, you may have to modify the filter rack to allow for additional surface area—just because one filter is the same size as another does not mean that the two filters have the same cfm and pressure drop ratings.

An air filter becomes more efficient at collecting particles as it "fills up"—however, a dirty filter also creates a greater pressure drop. The increase in pressure drop, or ESP, causes a reduction in the air flow. If the air flow becomes too low, the

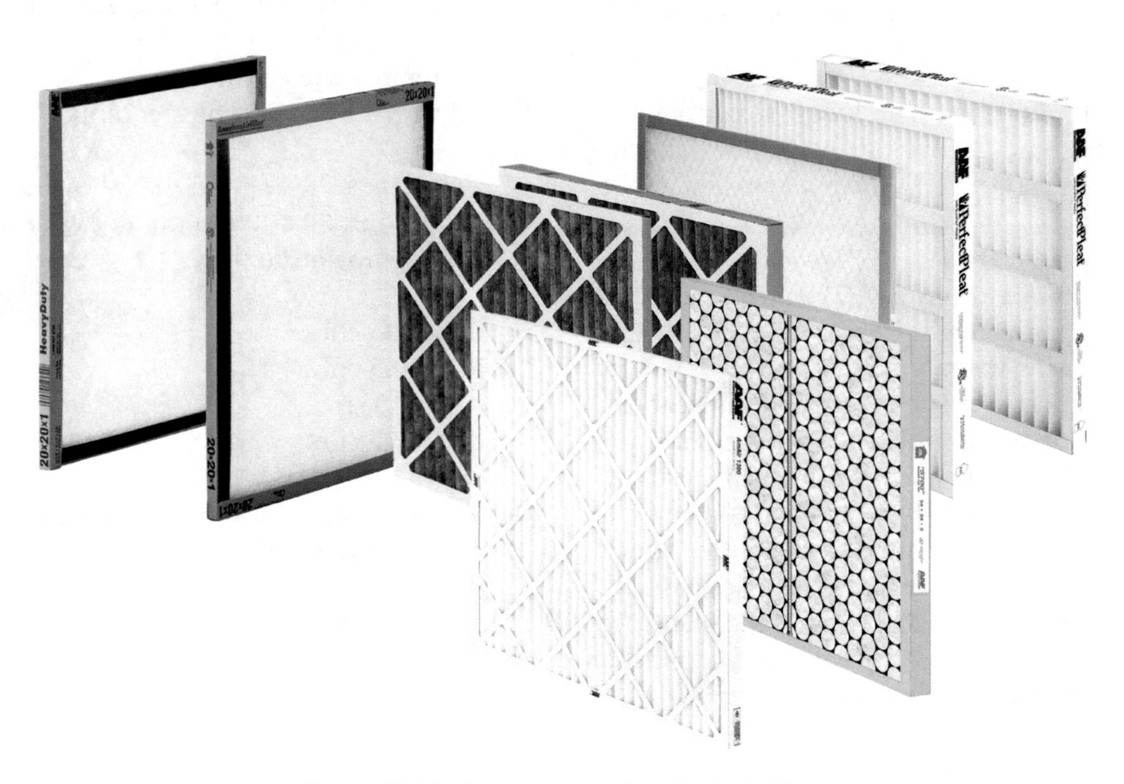

Figure 5-10. Assortment of typical air filters

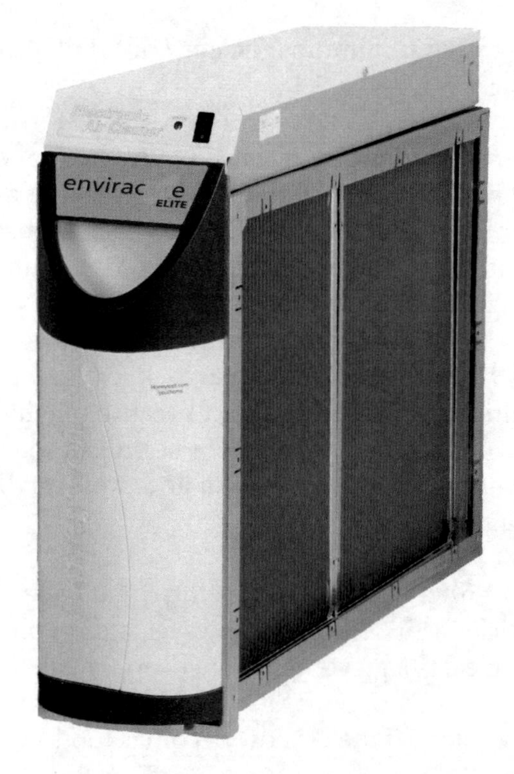

Figure 5-11. Typical electronic air cleaner (EAC)

heating and cooling capacity of the equipment will suffer, and the equipment may be damaged. *It is important to check and replace filters at regular intervals.*

An *electronic air cleaner* (EAC) works by electrically charging particles as they pass through ionizing wires, causing them to be attracted to oppositely charged collector plates. Electronic air cleaners like the one shown in Figure 5-11 can remove very small particles and are often beneficial for people with allergies. Pre-filters should be used with electronic air cleaners to avoid the "bug zapper" effect caused by large particles passing through the high-voltage wires. If the plates are not cleaned regularly, they will reach capacity and allow everything to pass straight through.

"Filter bypass" is a significant problem in many systems. Needless to say, air that does not pass through the filter cannot be cleaned. It is very important to make sure that all filters are put in place, and that they fit into their racks properly. Gaps between filters or around the ends allow air to pass unfiltered. □

REVIEW QUESTIONS

1. Too much air flow can cause __________.

   a. cycling on the limit switch
   b. increased energy consumption
   c. the furnace to shut down
   d. the heat exchanger to overheat

2. Before using the temperature rise method to calculate air flow for a gas furnace, you must first __________.

   a. clean the fan blade
   b. make sure that the furnace has not been running for at least 30 minutes
   c. remove the air filter
   d. verify the proper burner output

3. Calculate air flow given the following conditions: 80,000 Btu input, 84% efficiency, 45°F temperature rise.

   a. 1,295 cfm
   b. 1,383 cfm
   c. 1,510 cfm
   d. 1,777 cfm

4. Which of the following is the best instrument to use for measuring ESP?

   a. Anemometer
   b. Balometer
   c. Diaphragm-type differential pressure gauge
   d. U-tube manometer

5. Based on Table 5-1, what is the air flow from a vertical unit operating at 230 V with the motor on the medium speed tap when the ESP measures 0.30 in. w.c.?

   a. 800 cfm
   b. 850 cfm
   c. 900 cfm
   d. 950 cfm

6. When using velocity pressure to calculate air flow, you must be sure to _________.

   a. face the Pitot tube downstream from the direction of air flow
   b. get an accurate temperature reading
   c. traverse the duct
   d. use a velocity gun

7. What is a proper air velocity in residential trunk lines?

   a. 400 ft/min
   b. 600 ft/min
   c. 800 ft/min
   d. 1,000 ft/min

8. What is a recommended air velocity across a disposable filter?

   a. 200 to 250 ft/min
   b. 400 to 500 ft/min
   c. 700 to 750 ft/min
   d. 900 to 1,000 ft/min

9. Proper *throw* from a diffuser depends primarily on the _________.

   a. face velocity
   b. free area of the diffuser
   c. height of the diffuser
   d. temperature of the air being delivered

10. Insulation is required on ductwork when ducts are run _________.

    a. horizontally between floor joists
    b. vertically between wall studs
    c. overhead in dropped ceilings
    d. through unconditioned spaces

11. What is the minimum recommended MERV rating for filters used in residential equipment?

    a. 2
    b. 4
    c. 6
    d. 8

12. When is an air filter most efficient at moving air?

    a. When it is brand new
    b. After the initial break-in period
    c. At 50% of its life
    d. Near the end of its life

13. When is an air filter most efficient at trapping dirt?

    a. When it is brand new
    b. After the initial break-in period
    c. At 50% of its life
    d. Near the end of its life

14. Electronic air cleaners work by _________.

    a. electromagnetic force
    b. ionizing particles
    c. stripping neutrons from the particles
    d. transistor reaction

15. The problem of "filter bypass" is critical because it allows _________.

   a. an increase in external static pressure
   b. a reduction in overall air flow
   c. air to leak from the ductwork
   d. unfiltered air to pass through the equipment

ILLUSTRATION CREDITS
FIGURE 5-2: EXTECH INSTRUMENTS
FIGURE 5-3: DWYER INSTRUMENTS, INC.
FIGURE 5-5: ALNOR
FIGURE 5-10: AMERICAN AIR FILTER INTERNATIONAL
FIGURE 5-11: HONEYWELL

CHAPTER 6

Troubleshooting

ELECTRICAL TROUBLESHOOTING

Since 75 to 80% of all service calls involve electrical components in one way or another, it is very important for HVACR technicians to have a good basic understanding of electrical systems. This Chapter does not attempt to cover the fundamentals of electricity, but it does examine some of the individual electrical components of gas and oil furnaces.

Generally speaking, electrical components fall into two main categories—loads and switches. A *load* is a device that uses electricity to do some type of useful work, such as turn a motor, generate heat, or turn on a light. A *switch* is a device that acts as a "gatekeeper" in an electric circuit—a *closed* switch allows electric current to pass, while an *open* switch prevents the flow of electric current (see Figure 6-1 on the next page). A switch can be activated manually, or by a change in temperature, pressure, liquid level, or any other variable that can be measured.

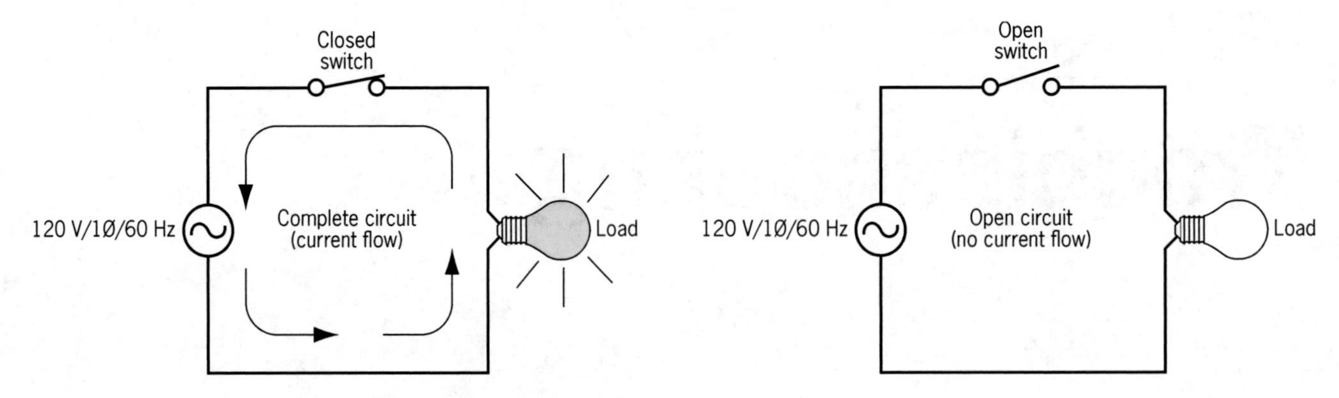

Figure 6-1. Loads and switches

The typical residential furnace has a 120-V power supply, with a disconnect switch located on the side of unit. The National Electrical Code (NEC) requires a means of disconnect within sight of the unit. New codes also require the installation of a green ground wire to the unit. In the past, metal conduit often was used as the ground path. Be sure to check local code requirements for variations. Consult the furnace manufacturer's literature to determine the minimum circuit ampacity or minimum circuit size for the equipment. If you arrive on a service call and the system is not operational, the first thing to check is the supply voltage. The disconnect may contain a fuse, which also should be checked if no power is present.

TESTING CIRCUITS

Circuits can be tested in two basic ways. When a circuit is energized, you can use a *voltage meter* to determine if the circuit is complete and power is being delivered to the load. The second method is to use an *ohmmeter* to measure resistance. By measuring resistance, you can determine if the path between two points is complete. Resistance testing can be done *only when the circuit is de-energized.* This safety measure not only protects the technician from possible injury, but also prevents damage to the meter and the equipment.

The components being tested must be isolated from the rest of the circuit. Whether you are using a voltage meter or an ohmmeter, be sure to select a meter setting that is *above* the expected voltage or resistance to prevent damage to the instrument. A digital *multimeter* like the one shown in Figure 6-2 is capable of measuring

Figure 6-2. Digital multimeter

both voltage and resistance, which eliminates the need for using separate meters.

Voltage testing

When you perform voltage tests to check circuits, you are working with live voltages. *Always* wear the proper personal protective equipment and use extreme caution to avoid contact with energized parts. Both loads and switches can be checked by measuring voltage. The voltage test for a load is done simply to ensure that the proper voltage is getting to the load. It is important to verify that the voltage applied to the load is within an acceptable tolerance of ±10% of the load's rated voltage. A load that has the proper voltage applied but does not operate generally has an open internal circuit or has failed mechanically.

Switches also can be checked with voltage tests, as shown in Figure 6-3 below. Remember that voltage is a measurement of potential difference, so when there is a difference from one side of the switch to the other—that is, when you get a voltage reading that is the same as or nearly the same as the applied voltage—it means that the switch is open. When there is no difference, you get a voltage reading of zero and it means that the switch is closed.

It is important to remember that these readings are accurate *only if the rest of the circuit is complete.* If a second switch is connected in series with the first switch and the second switch is open, then the first switch will read 0 V in both the open and closed positions (see Figure 6-4 on the next page). A partial voltage reading across a switch is an indication that the contact points of the switch are pitted and not making full contact. This usually is a sign that the switch or relay needs to be replaced.

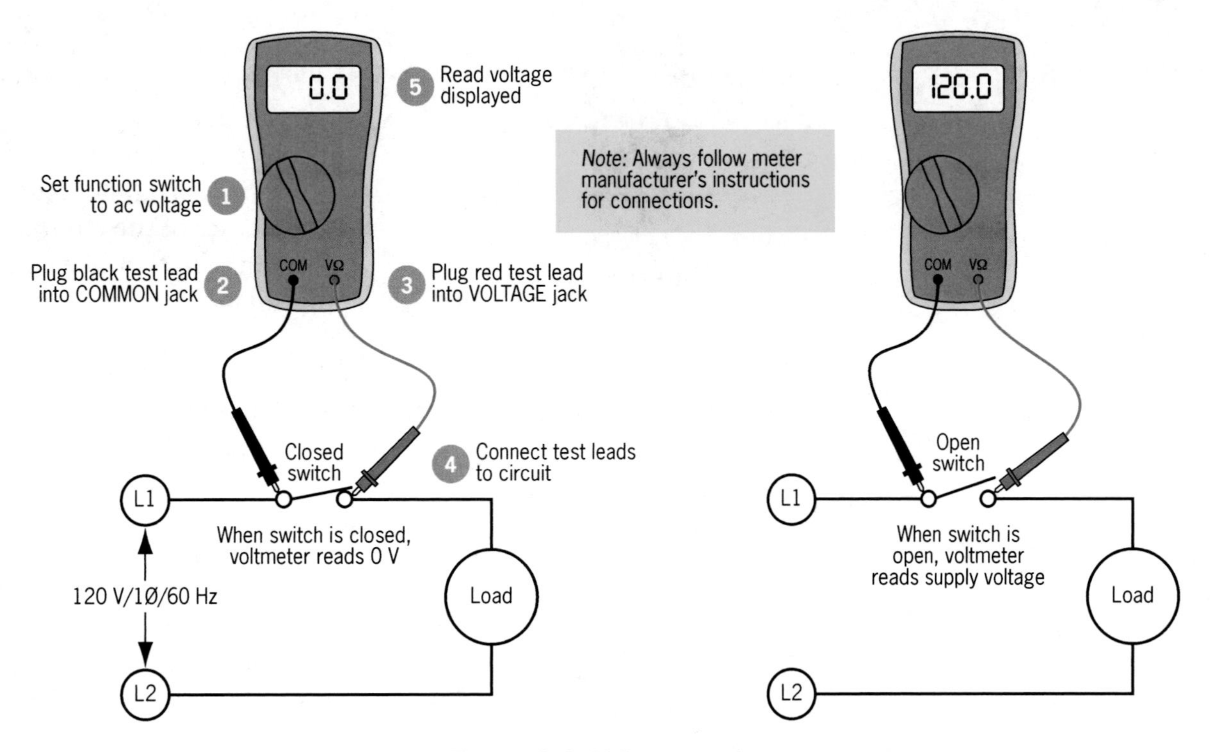

Figure 6-3. Voltage testing

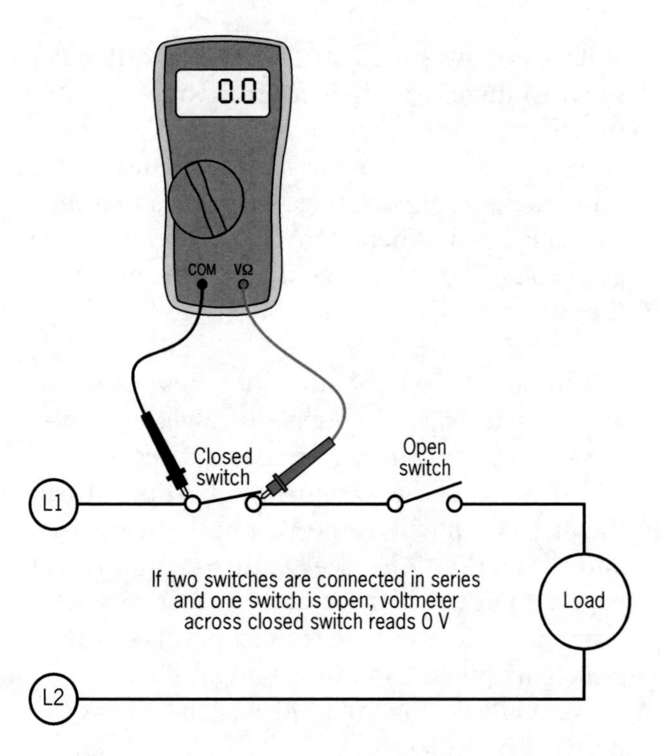

Figure 6-4. Switches connected in series

Because measuring voltages across switches is not always accurate, a better way to check the switches in an energized circuit is the "hopscotch" method. This technique consists of attaching one meter probe to the neutral or L2 line, and moving the other probe from point to point across the circuit (see Figure 6-5). A voltage reading on both sides of the circuit indicates a closed switch, while a reading on one side but not the other indicates an open switch. This method is very beneficial in checking a control or safety circuit that has multiple switches connected in series. In order for power to be delivered to the load, all of the switches in the series must be closed.

Resistance testing

When you check loads and switches by measuring resistance, make sure that the power to the unit is turned off and that the components being tested are isolated (see Figure 6-6). All functional loads have some resistance. Typically, the larger the load is, the lower the resistance will be. To determine the "normal" resistance of a component, you can consult the manufacturer's specification sheet, or compare the resistance of the component under test with that of a similar component that is known to be good.

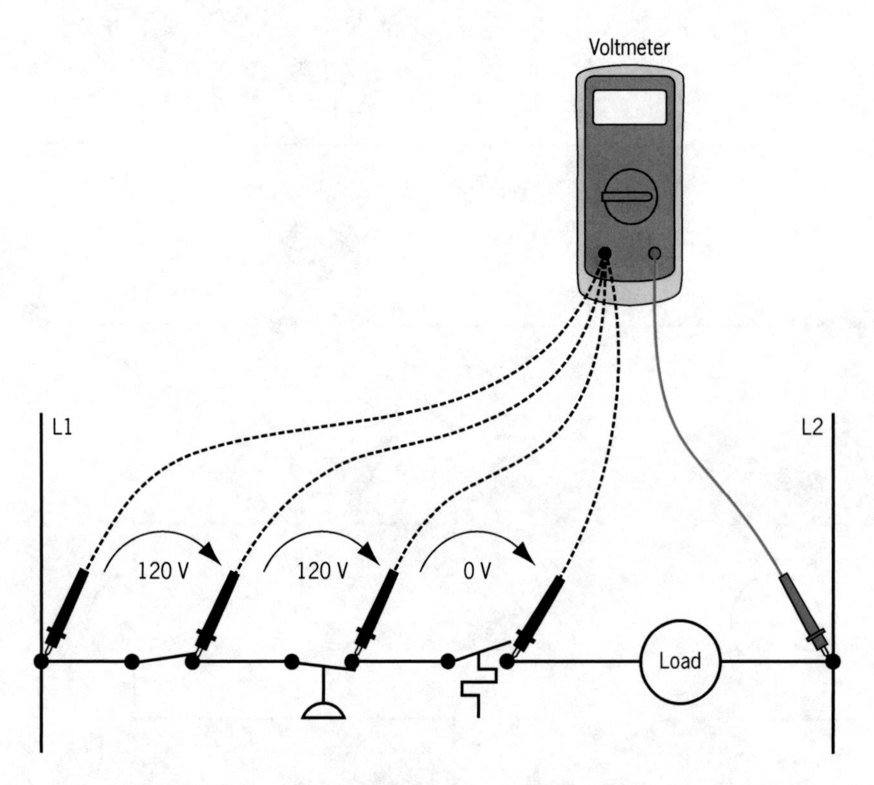

Figure 6-5. Using the "hopscotch" method of troubleshooting

A resistance reading of 0 Ω in a load indicates the presence of a short circuit—a condition that generally opens a fuse or trips a circuit breaker. An infinite ("OL") resistance reading is an indication of an open circuit or an incomplete path. Be aware that you will also get an infinite reading if a motor has an open internal overload.

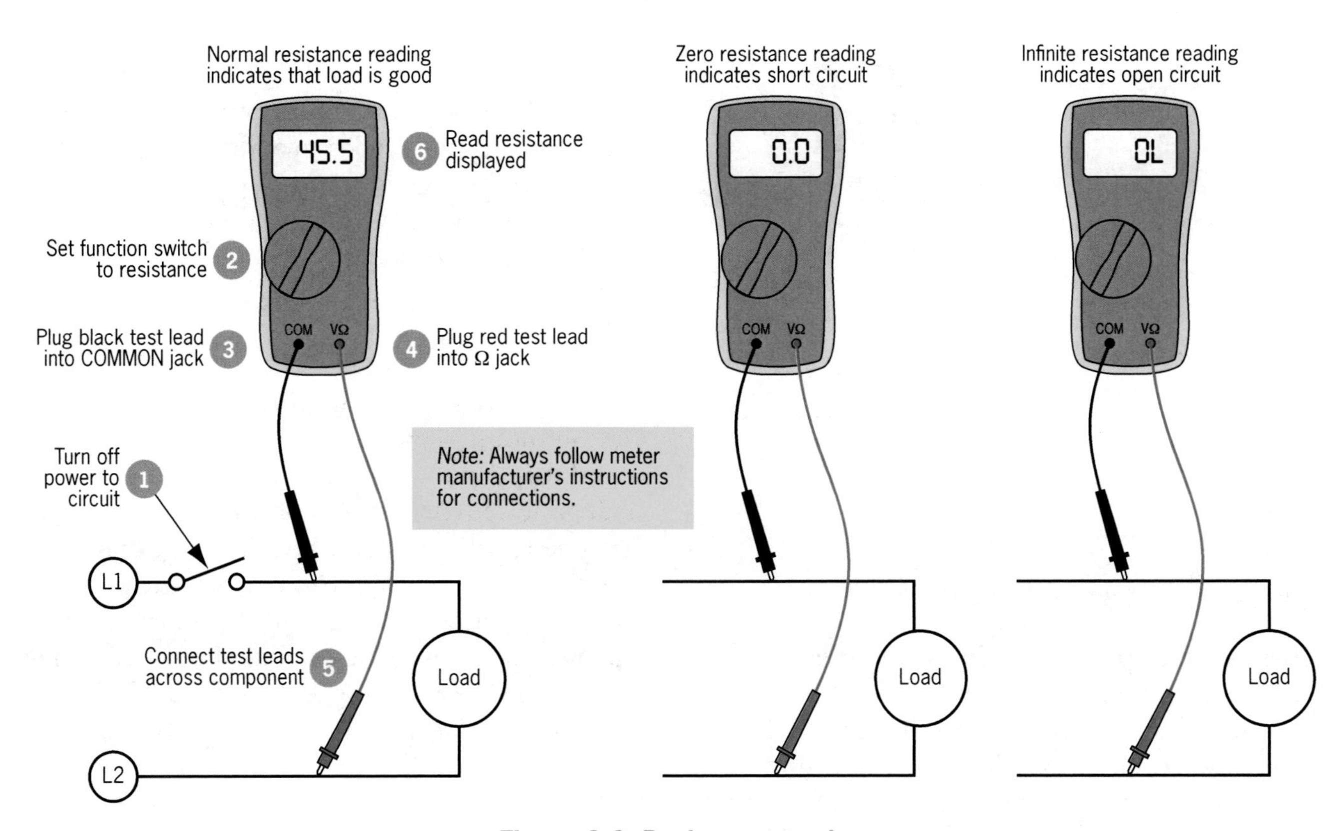

Figure 6-6. Resistance testing

When you use an ohmmeter to check for resistance across a switch, the following outcomes are possible (see Figure 6-7):

- A measurement of 0 Ω indicates that the switch is closed.
- An infinite resistance reading indicates that the switch is open.
- A closed switch that produces a measurable resistance reading probably has pitted contacts and should be replaced.

A *continuity test* is a simple resistance test that shows whether there is a connection between two points. Many meters provide an audible signal to indicate continuity (see Figure 6-8 on the next page). Continuity generally is confirmed as long as the resistance between the two points is below about 100 Ω.

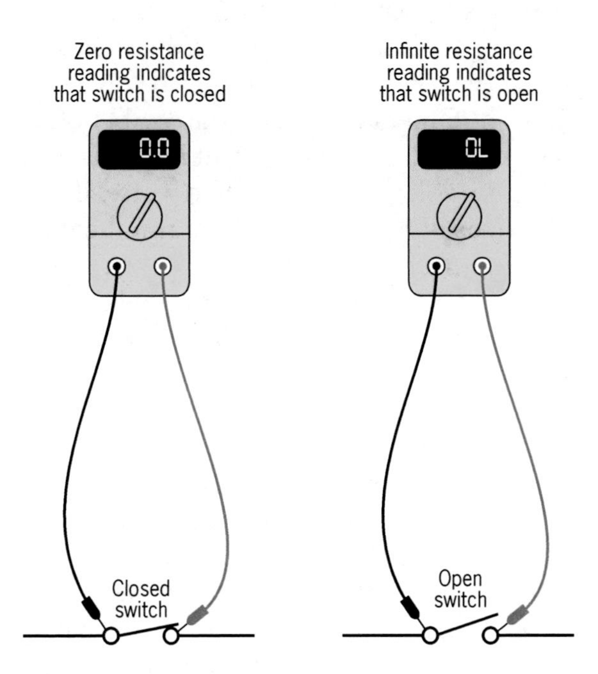

Figure 6-7. Checking resistance across switches

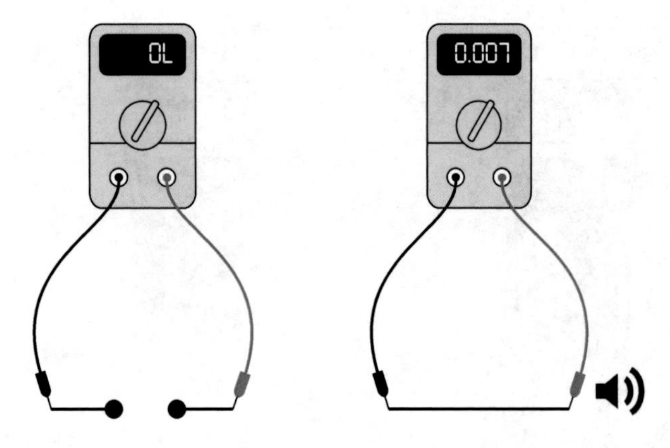

Figure 6-8. Continuity testing

GENERAL ELECTRICAL CHECKS

Before start-up and at periodic intervals when planned maintenance is performed, the technician should do a visual inspection of all electrical connections. Signs of overheating or corrosion at any of the connections are evidence of a problem. Overheating is often caused by a loose or poor connection. The bad connection creates resistance, which in turn causes a voltage drop that generates heat. Corrosion often is caused by moisture, but also may be the result of the equipment overheating over a period of time. A corroded wire acts as an insulator and makes the voltage drop problem worse. If wires show signs of heat or corrosion, trim them back to expose clean wire and then reconnect them. Check all terminals for tightness and broken wires to ensure that nothing has worked loose due to vibration.

Check the voltage to the unit with no load operating. This will help you determine what the power company is supplying to the building. Next, check the voltage during start-up to make sure that it does not drop below the 10% allowable tolerance of the equipment. Due to the *inrush* current (the temporary high current required to start motors and energize coils), the highest current draw—and therefore the greatest voltage drop—should occur during start-up. After the load is running, take an additional voltage reading. If the voltage has dropped more than 2% from the no-load voltage, a problem with conductors or wiring may be indicated.

WIRING DIAGRAMS

A wiring diagram is an important tool for troubleshooting an electrical system. It provides a "road map" that shows how all the components are connected. It is important for technicians to become familiar with the use of wiring diagrams. If a diagram is missing, you may have to trace the wiring and sketch out your own diagram for the piece of equipment. Making the sketch is like connecting the dots—once enough of the dots have been connected, the picture starts to become

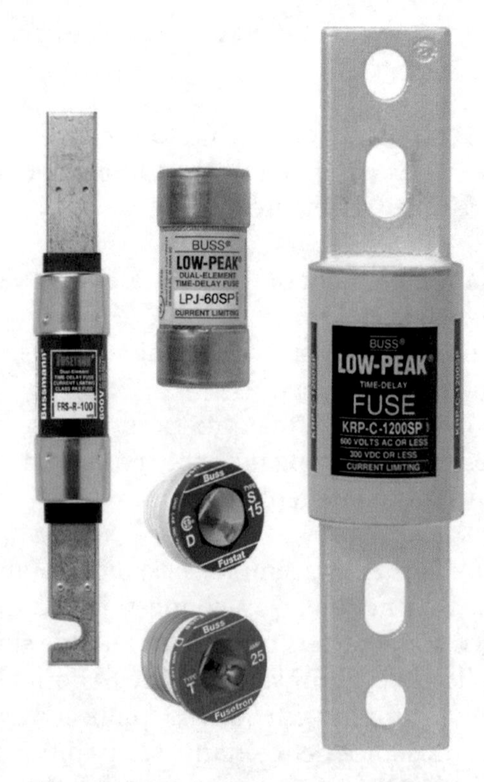

Figure 6-10. Various types of fuses

clear. Figure 6-9 on pages 72 and 73 shows a typical ladder diagram for a furnace. More information about how to read electrical schematics is available in the *Core Essentials* book in this series.

TROUBLESHOOTING COMPONENTS

Fuses

A *fuse* is a safety device that opens (breaks the circuit) in the event of a high current draw. Fuses come in a variety of types for different applications (see Figure 6-10). Time-delay fuses are often used for HVAC work because they allow a higher current draw for a short period of time (as required by the inrush current of a motor during start-up). The *current rating* of a fuse is the maximum current that can be drawn for an extended period of time. The *voltage rating* of the fuse is the maximum voltage that can be applied to the fuse.

To test a fuse for continuity, remove or isolate the fuse and check the resistance across the fuse (see Figure 6-11). A resistance reading of 0 Ω indicates that the fuse is closed and still good. An infinite resistance ("OL") across the fuse indicates an open fuse that needs to be replaced. To test a fuse with a voltage meter, first check the incoming line voltage. The reading should be the same as the applied voltage of the system. Then check the load side of the fuse to neutral. If the applied voltage is read, the fuse is good. If there is no reading, the fuse is defective and must be replaced.

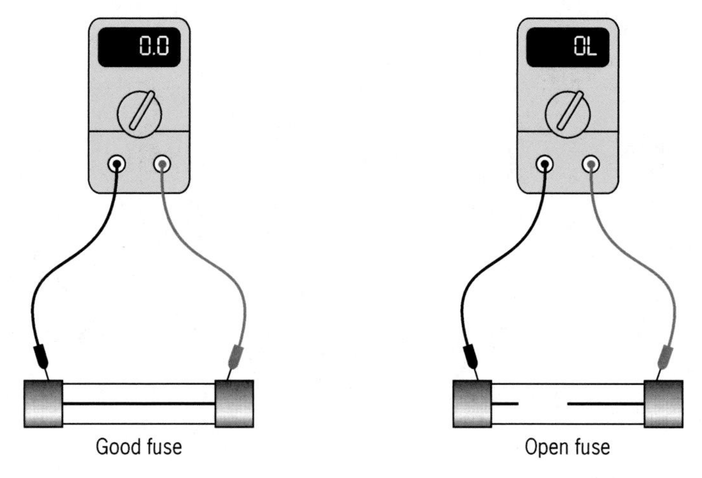

Figure 6-11. Checking resistance across fuses

Transformers

The job of a *transformer* is to "step down" (decrease) or "step up" (increase) voltage. An example of the former is a transformer used to convert 120-V supply voltage to 24-V control voltage (see Figure 6-12). An example of the latter is the ignition transformer on an oil burner (see Figure 6-13 on page 74).

Voltage is induced into the secondary side of a transformer by a magnetic field produced by the primary coil. The two sides of a transformer are not connected electrically, so checking a transformer requires testing both sides. A transformer is classified by its primary and

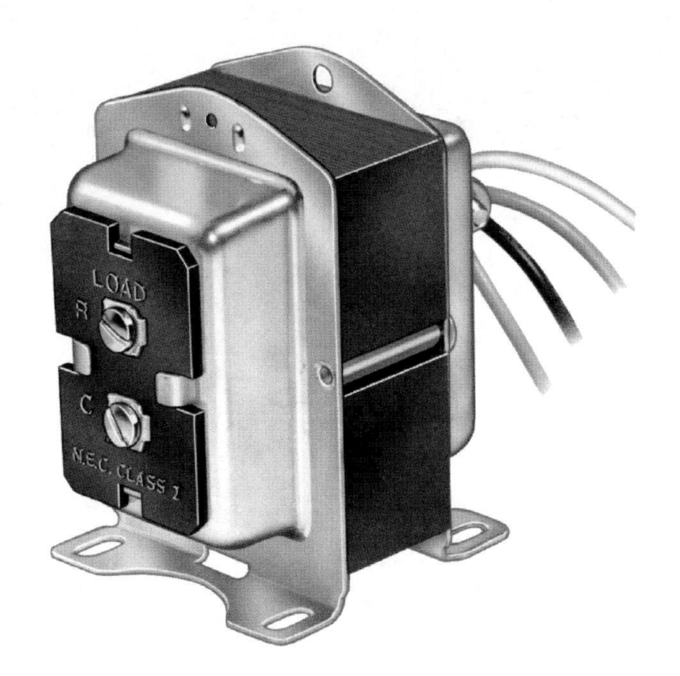

Figure 6-12. Typical transformer

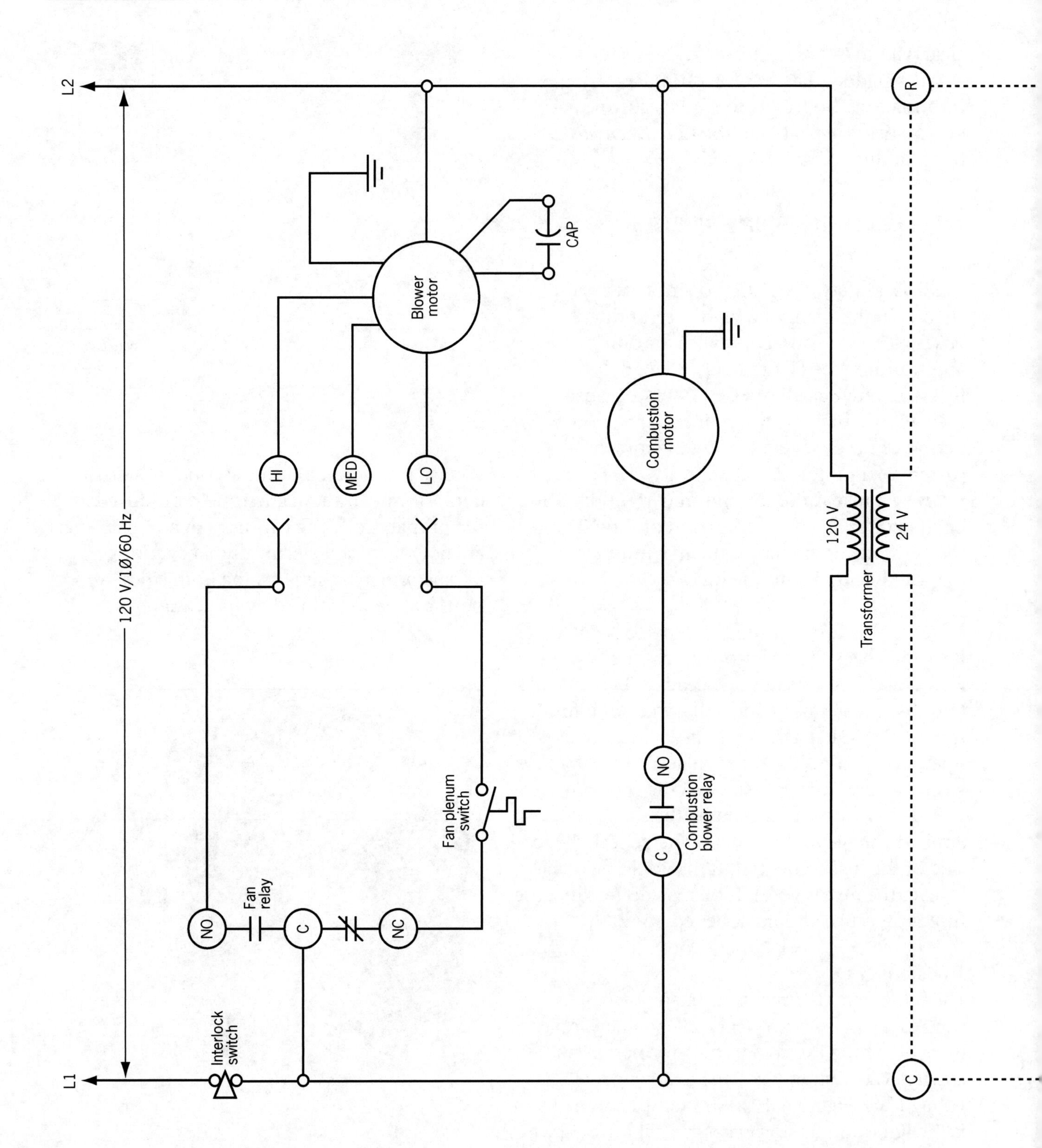

Figure 6-9. Schematic diagram of gas furnace with intermittent pilot or spark ignition

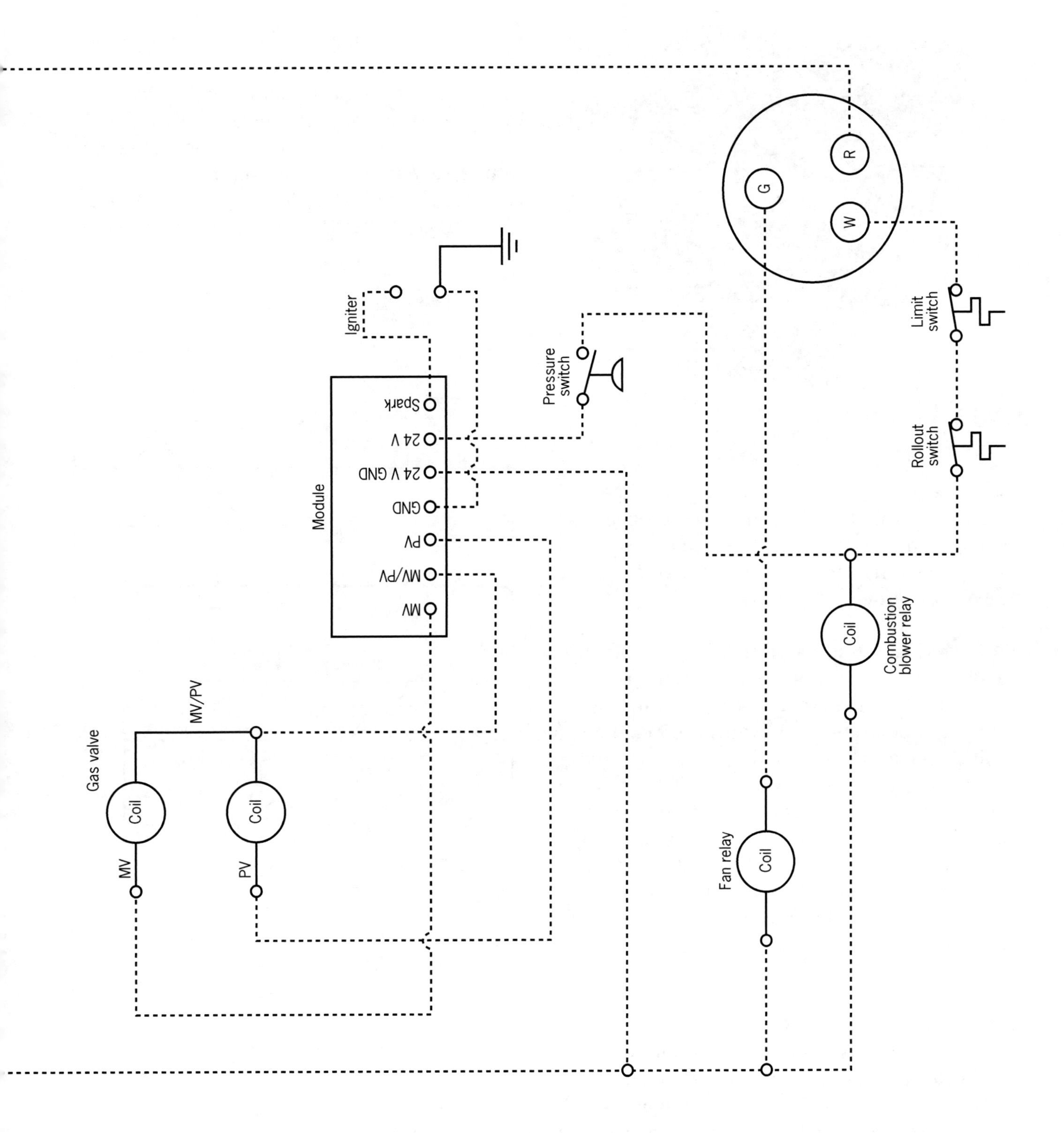

Figure 6-9. Schematic diagram of gas furnace with intermittent pilot or spark ignition (continued)

Figure 6-13. Ignition transformer

secondary voltages, and by its VA (volt-ampere) rating. The VA rating divided by the secondary voltage indicates the current available at the secondary. A transformer must be sized to handle the inrush current of the system loads.

To do a voltage test, start by checking the voltage at the secondary side of the transformer (see Figure 6-14). If the proper voltage is available at the secondary, the transformer is good. If no voltage is read at the secondary, then check the voltage at the primary. If there is voltage at the primary but not at the secondary, the transformer is bad. To do a resistance test, measure the resistance at the primary and secondary sides of the transformer. Both sides should read some resistance on the windings. A zero or infinite resistance indicates a bad transformer.

Motors

Motors are used in HVAC systems for blowers, induced-draft fans, compressors, condenser fans, oil burner pumps, and more. There are many types of motors, but all generate a magnetic field to create rotation. To do a voltage test on a motor, first check the power to the motor. If the proper voltage is being supplied, the motor should run. If the motor fails to operate, further checks need to be done before the motor is condemned. An open overload, a bad capacitor, or a bad start relay may prevent a good motor from operating. If the system motor is an ECM (electronically commutated motor), follow the manufacturer's instructions for checking the motor and module.

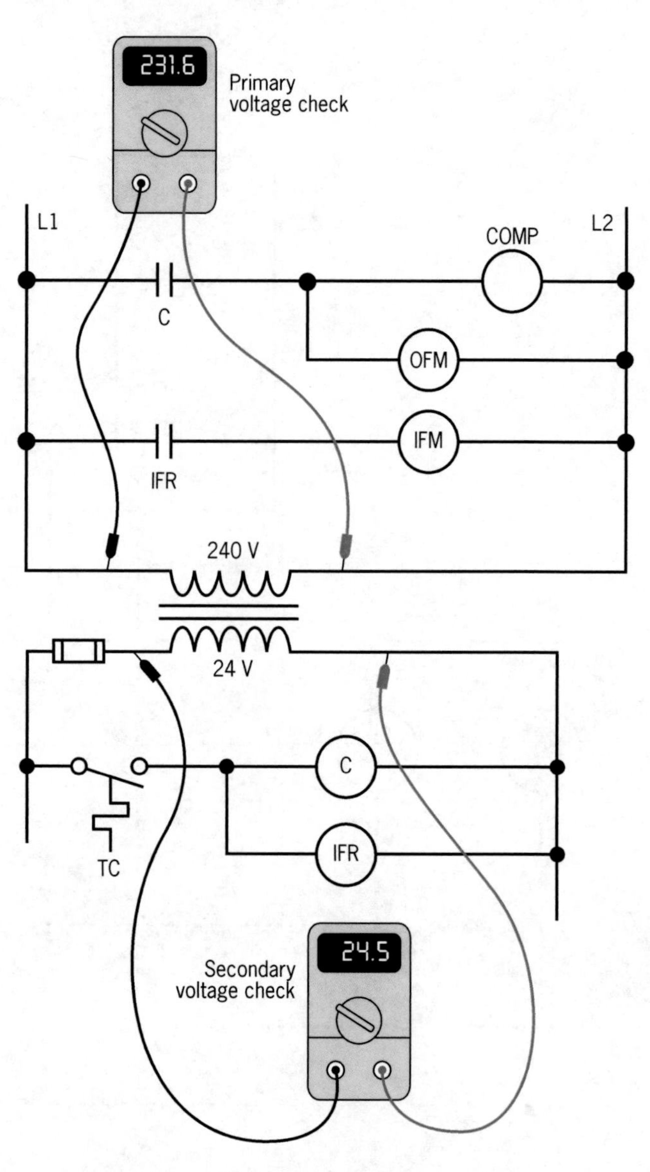

Figure 6-14. Checking voltages at control transformer (typical wiring diagram)

Check the motor windings for resistance (see Figure 6-15). Since the motor is a load, it should exhibit some measure of resistance. A resistance reading of 0 Ω indicates a shorted motor winding. An infinite resistance indicates an open circuit, which could be an open winding or an open overload. Be sure to check both the start and run windings. In addition, measure the resistance from each winding lead to the motor shell. You should measure infinite resistance. If you measure any amount of resistance, the motor is shorted to ground. Figure 6-16 shows how to test a permanent split-capacitor (PSC) motor.

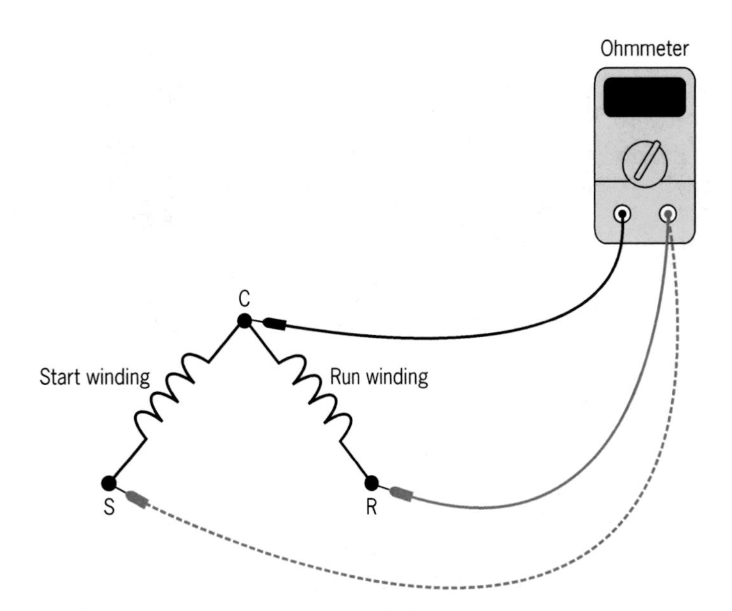

Figure 6-15. Testing PSC motor windings for resistance

Many of the motors used in the HVAC industry have internal overload protectors. If the overload is open, the motor will not run even if voltage is applied. If the motor feels warm or hot to the touch but does not run, it is a good indication that the overload is open. It can take several minutes for an open overload to reset. If the overload is open, a resistance test from run to common will yield an infinite ("OL") reading, as will a test from start to common. However, a test from run to start should measure some resistance, since the run and start terminals are wired together at the common point (see Figure 6-17 on the next page).

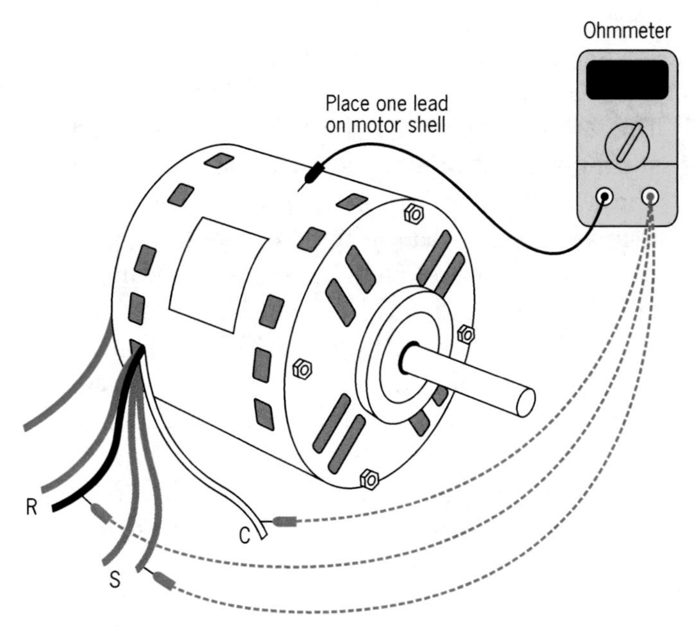

Figure 6-16. Testing PSC motor for grounds

A motor is an *inductive* load, which means that it creates a phase shift between the voltage and the current. Known as the *power factor*, this phase shift increases the motor's current draw. To correct for this effect and improve the power factor, capacitors are added to the motor circuit. A capacitor also causes a phase shift, but one that acts in the opposite direction from the load, thus *reducing* the amount of current drawn by the motor. The addition of a capacitor in the start winding also improves the motor's starting torque.

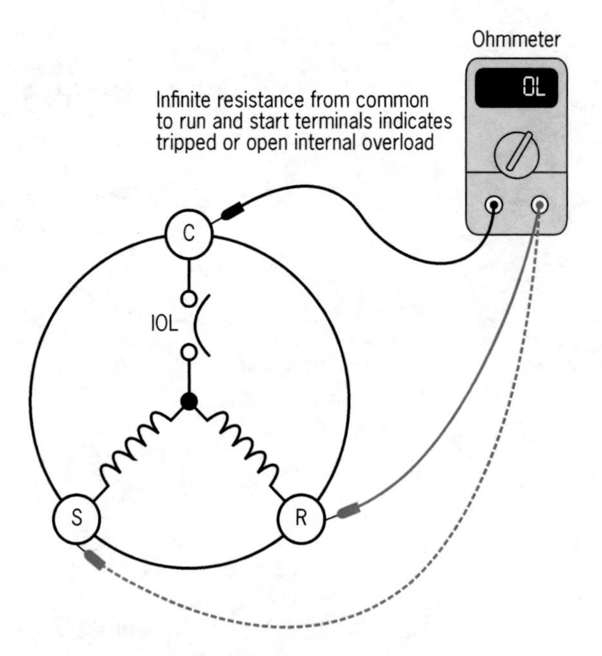

Figure 6-17. Checking internal overload of PSC motor for resistance

There are two types of motor capacitors—start capacitors and run capacitors (see Figure 6-18). The *start capacitor* is only in the circuit for a very short period of time while the motor is starting. The start capacitor's primary function is to increase the starting torque of the motor. Once the motor is up to speed, the start capacitor is taken out of the circuit by a centrifugal switch or a start relay. The *run capacitor*, which is filled with oil as a means of dissipating heat, is connected in the circuit whenever the motor is running. In a single-phase motor, the run capacitor is used to provide a second phase (or "phase shift") that produces a rotating magnetic field. Run capacitors are required for starting PSC motors.

You can use an analog ohmmeter to perform a basic check of a capacitor (see Figure 6-19). When you place the meter probes across the capacitor terminals, the needle should swing quickly toward zero and then more slowly come back toward infinite. A zero or constant resistance reading indicates a shorted capacitor. An infinite reading indicates an open capacitor. It is important to know your meter—if the meter is not set on the correct scale, the meter can react so fast that it does not appear to move. Check each terminal of the capacitor to the metal case. A zero resistance reading between any of the terminals and the case means a shorted capacitor. To do a thorough test of the capacitor, use a digital capacitance meter that measures the microfarad value directly (see Figure 6-20).

Figure 6-18. Typical motor capacitors

Although this Chapter concentrates primarily on electrical tests, troubleshooting a motor frequently requires some mechanical checks as well. Inspect the motor shaft to make sure that it turns freely. Any up-and-down movement or in-and-out movement (end play) indicates bearing wear. Some motors may require periodic lubrication. Sleeve bearings generally are used on smaller, lighter-duty motors because of their quiet operation. These motors usually require the addition of a lightweight, non-detergent oil. Larger or heavier-duty motors commonly use ball bearings. These bearings may require the periodic addition of grease. Some motors are considered permanently lubricated and do not need any additional lubrication. Always take the time to check the manufacturer's lubrication requirements.

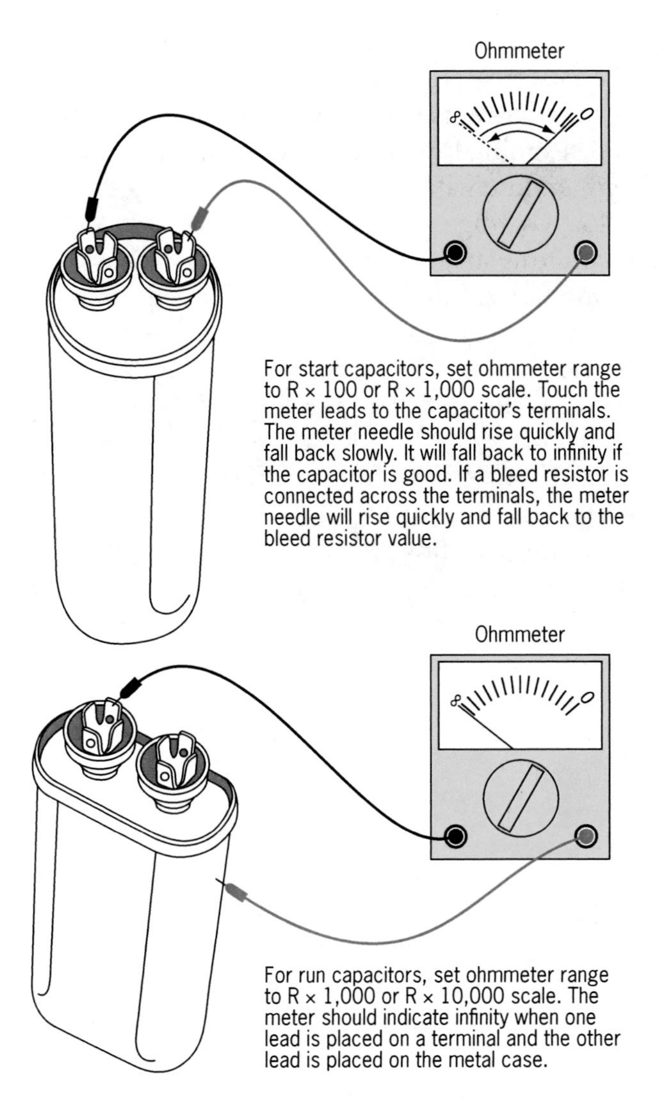

Figure 6-19. Checking capacitors for resistance

Figure 6-20. Capacitance meter

If a motor does need to be replaced, it is important to gather the following information:

- horsepower
- voltage
- frame size
- rotation
- shaft size
- ECM programming requirements.

The manufacturer's name and model number also can be very helpful as a cross-reference when you are selecting a new motor.

Relays and contactors

Relays and *contactors* are circuit control devices (a typical contactor is shown in Figure 6-21 below). They contain sets of switches that are controlled by electrical signals. Relays and contactors perform basically the same function, except that contactors operate larger loads. Both may have multiple contacts. A relay generally is used to allow a low-voltage control system to operate line-voltage loads. The relay has separate electric circuits contained in one physical package. The relay *coil* is an inductive load that creates the magnetic field to move the switches. If a voltage test shows that the proper voltage is being applied to the coil, the switches should change position. A resistance test also can be done on the coil—since it acts as a load, it should exhibit some measure of resistance.

A relay may have any number of switches, which can be *normally open* (NO), *normally closed* (NC), or any combination of the two. The "normal" position is the position of the switch with no

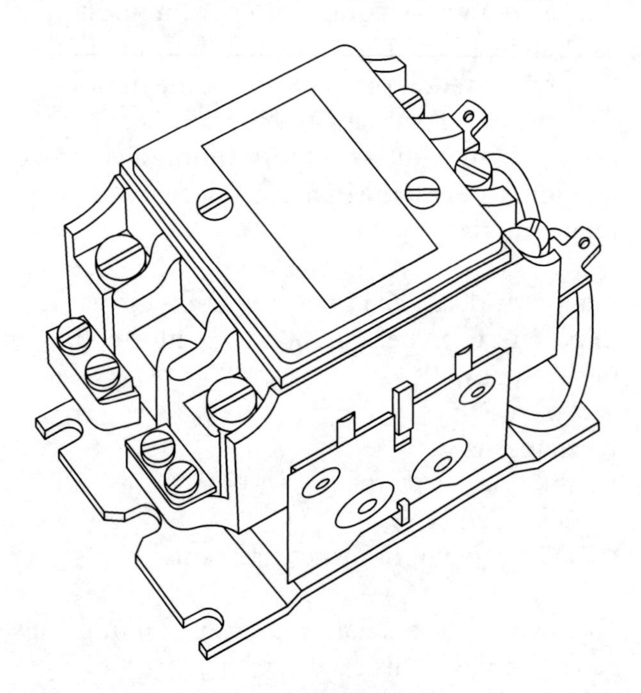

Figure 6-21. Typical contactor

power applied to the coil. To check the relay's switches, measure the voltage at the input of the relay to the neutral side. If power is present at the line (input) side of the relay and the coil is energized, you should obtain a voltage reading on the load (output) side of the relay, indicating a normally open switch. If the switch fails to close, the coil or the contacts are bad and the relay will need to be replaced. When you replace a defective relay with a new one, it is important to match the number of poles (switches), the current rating of the switches, and the coil voltage.

Thermostats

The *thermostat* is the primary control for starting and stopping furnace operation based on the temperature within the conditioned space. Low-voltage wiring is run between the furnace and the thermostat. Per the NEC, low-voltage wiring cannot be run in the same conduit with line-voltage wiring. Figure 6-22 shows the wiring connections to the thermostat. Low-voltage wiring is normally color-coded as follows:

- red: 24-V power from the transformer
- green: fan circuit
- yellow: cooling circuit
- white: heating circuit
- blue: 24-V common from the transformer.

The thermostat basically consists of two parts. The main body of the thermostat contains the temperature-activated switches. The *subbase* contains the function switches (such as the HEAT-OFF-COOL and FAN-AUTO switches). Since the thermostat is essentially a series of switches, it should be checked as switches. Check the terminal block at the furnace to determine whether the thermostat is providing the appropriate voltage. If a switch does not close or if a wire breaks on the thermostat, then that particular function will not operate. If a switch shorts, it will provide a constant voltage and that function will continue to operate.

On start-up and at regular service intervals, the thermostat should be run through its cycles to ensure that all functions are working properly. First turn the fan switch to the "ON" position and verify that the fan starts. This will show both that power is being supplied to the thermostat, and that the fan function is working. Next set the thermostat to the heating mode and adjust the setpoint to call for heat. Some time delay may be involved, but verify that the heat starts. Continue to cooling if applicable.

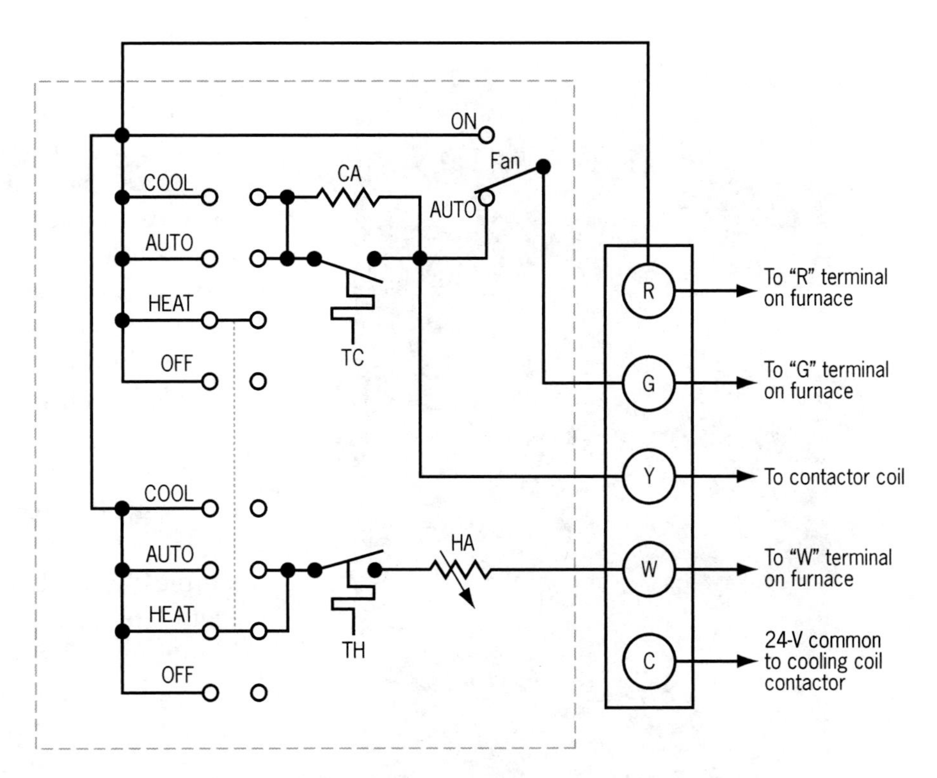

Figure 6-22. Typical thermostat wiring diagram

Gas valves

Let's assume that when the ignition cycle starts, the draft fan comes on and the hot-surface igniter or spark igniter powers up, but the main burner fails to light. This set of conditions will lead you back to the gas valve and gas supply. Before checking the gas valve, make sure that all manual shutoff valves are open—if possible, verify that there is gas coming to the house by checking another appliance. If piping work has been done, air will need to be vented, which may requirea few cycles of the unit.

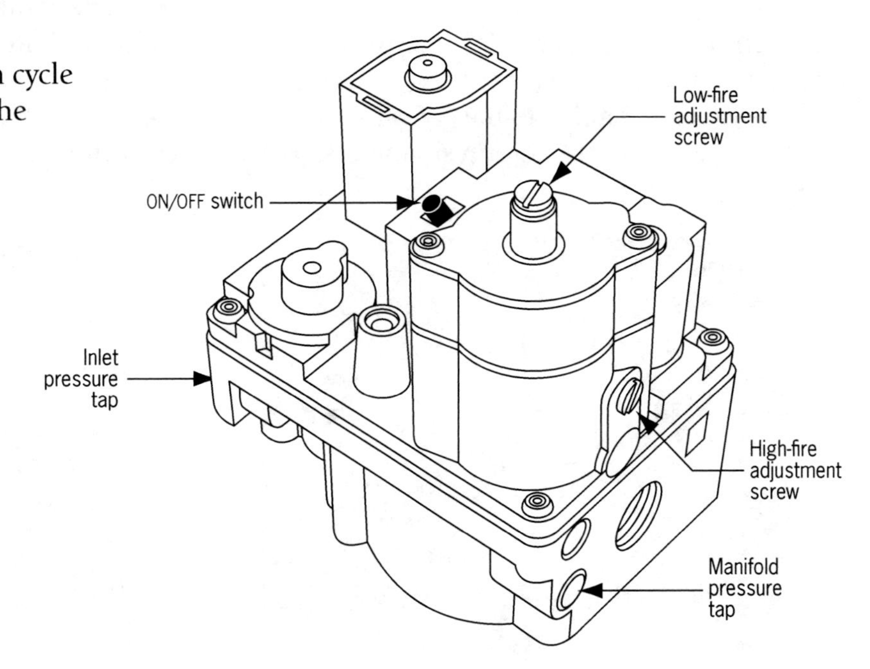

Figure 6-23. Gas valve

The gas valve (see Figure 6-23) is a load that is made up of two

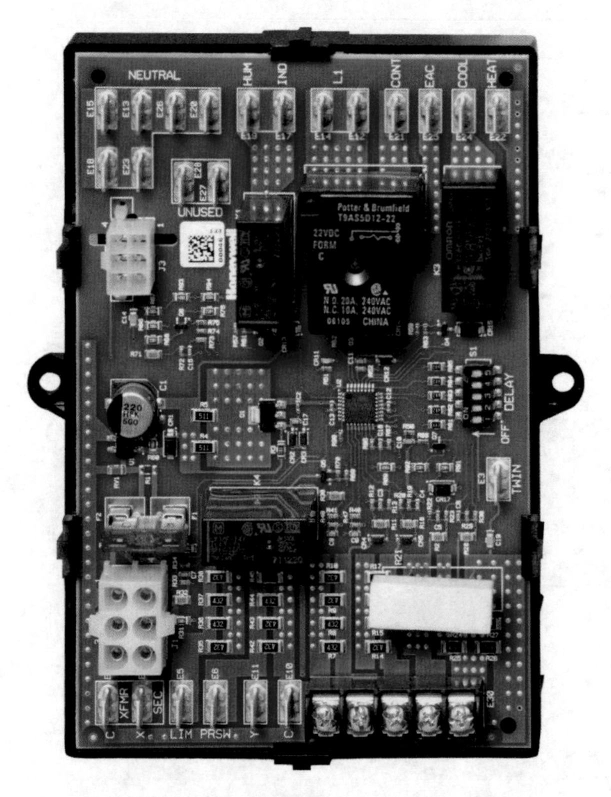

Figure 6-24. Electronic control board

magnetic coils. When energized, the coils open the valve. Start by checking to see if the proper voltage is being applied to the gas valve. If the valve fails to open, check for resistance. Be aware that in a standing-pilot gas valve, the pilot valve is operated on a millivolt power supply. When you test the coil for resistance, you must remove the thermocouple that supplies the power and take the measurement inside the socket. The resistance will be low compared to the 24-V coil, due to the very low operating voltage. It is important to remember that the valve may work properly electrically, but have a gas supply problem—such as a plugged inlet strainer, for example.

Electronic control boards

Electronic control boards like the one shown in Figure 6-24 may look complicated, but there are simple checks that you can use to troubleshoot them. The first step is to determine if the proper inputs are being fed to the control board. They generally include some type of power supply, and some type of control signal (such as a call for heat, indicated by 24 V on the "W" terminal).

If the inputs are correct, the next step is to check the circuit board fuses or circuit breakers, and then check the outputs (such as power to the induced-draft fan and hot-surface igniter). If the inputs received are producing the correct outputs, then you can conclude that the board is working correctly.

Figure 6-25. Hot-surface igniter

Figure 6-26. Door switch

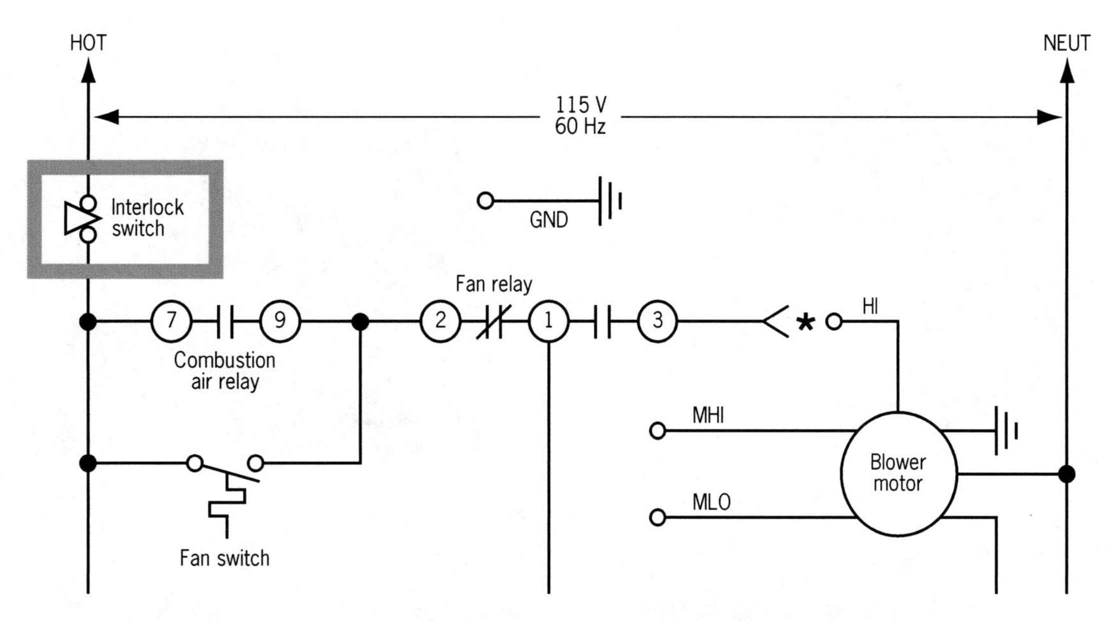

Figure 6-27. Furnace wiring diagram (door switch highlighted in green)

Many electronic control boards also have their own built-in diagnostic systems, which provide a series of blinking LEDs to lead the troubleshooter in the right direction. A visual inspection of the board often reveals burn marks or signs of corrosion. Corrosion may indicate another problem—a leak or condensation—that needs to be corrected.

Hot-surface igniters

The *hot-surface igniter* (see Figure 6-25) is a heating element that generally operates at 120 V (although some igniters do operate at 24 V). Most ignition modules perform an internal check of the igniter by making sure that the resistance is within an acceptable range. It is very important that you do *not* handle the heating element part of the igniter. Oils from your fingers can create hot spots and damage the igniter. Check to be sure that the proper voltage is being applied to the igniter. When you check the resistance at the igniter's leads, you should get a resistance reading. An infinite reading means that the igniter is open and needs to be replaced.

Switches

A residential furnace contains many switches. They vary by type and brand, but all switches can be checked in the same manner. Determining whether a switch should be open or closed requires a closer look at the variable activating that particular switch.

Door switches. The *door switch* (see Figure 6-26) acts as the first safety device on a modern furnace. If the furnace door is open, the door switch opens and interrupts power to the furnace so that nothing will work. You can test this switch electrically like any other switch to determine if it is open or closed. Be aware, however, that the panel or bracket holding the switch often gets bent. When this happens, the switch may not close even though the door is in place. If you can manually push in the switch and it closes, the switch is fine. *Do not bypass the switch for any reason*—make corrections to the panel or switch mounting to make it work properly. Figure 6-27 above highlights the door switch in a typical furnace wiring diagram.

Figure 6-28. Differential pressure switch

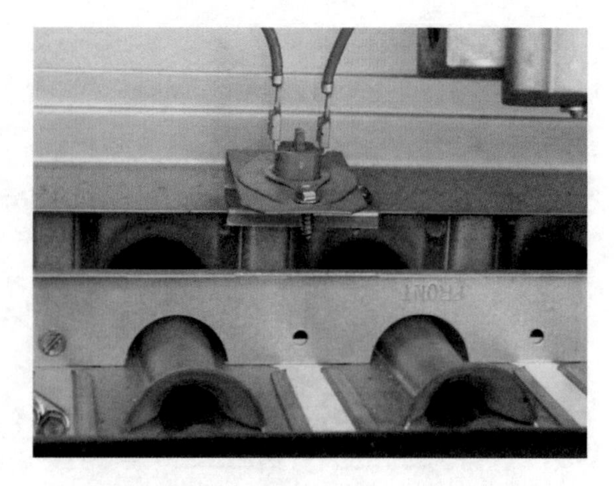

Figure 6-30. Rollout switch

Differential pressure switches. A *differential pressure switch* (see Figure 6-28) is commonly used to prove air flow for the induced-draft fan. To verify that a differential pressure switch is working properly, you must measure the pressure with a manometer. Remove the tubes to the switch and connect the meter into the switch where the lines were removed. A dual-port electronic manometer like the handheld model pictured in Figure 6-29 is the easiest instrument to use for this measurement.

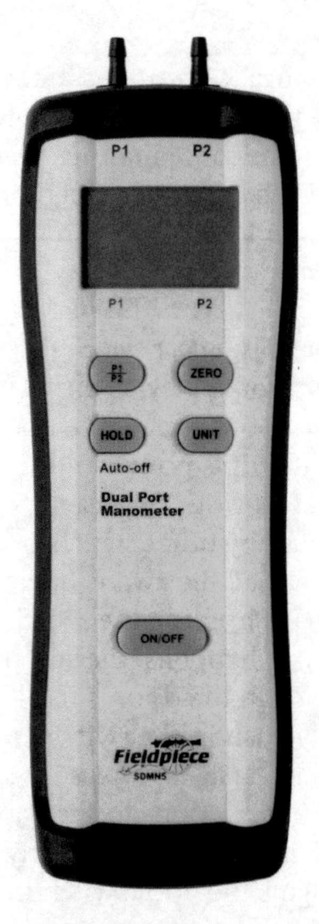

Figure 6-29. Dual-port manometer

Temperature limit switches. *Temperature limit switches* are placed in various locations within a furnace to provide protection from different types of abnormal operating conditions. Some temperature limit switches are "auto-reset" types, and others must be reset manually.

When a temperature limit switch opens, it means that something is not operating properly. It is important for the technician to investigate the cause and correct it, instead of simply resetting the switch. Some types of temperature limit switches are listed below:

- A *high limit switch* protects the furnace from damage if the air inside the supply plenum gets too hot.
- A *rollout switch* opens in the event of an abnormal flame pattern (see Figure 6-30).

Figure 6-31. Vent spill switch

- A *vent spill switch* turns off the fuel supply if combustion gases are escaping from the flue or chimney (see Figure 6-31).
- A *blower switch* shuts off the furnace in a downflow installation if the blower compartment gets too warm before the fan comes on. □

REVIEW QUESTIONS

1. The disconnect for a typical residential furnace is generally located __________.

   a. in the breaker box
   b. inside the front panel
   c. on the side of the unit
   d. on the wall behind the unit

2. When you arrive to service a unit and nothing is operating, the first thing you should check is the __________.

   a. door switch
   b. gas valve
   c. power supply
   d. thermostat

3. Which of the following should be checked *only* with the circuit de-energized?

   a. Current
   b. Resistance
   c. Voltage
   d. All of the above

4. What is an acceptable voltage tolerance for most loads?

   a. ±5%
   b. ±10%
   c. ±5 V
   d. ±10 V

5. A voltage reading across a switch that is the same or nearly the same as the applied voltage indicates that the switch is __________.

   a. open
   b. defective, probably due to corroded contacts
   c. wired in series with another switch
   d. all of the above

6. What type of reading should you get when you measure the resistance across a closed switch?

   a. 0 Ω
   b. Some measurable amount of resistance
   c. Infinite resistance
   d. OL

7. What type of reading should you get when you measure the resistance of a good load?

   a. 0 Ω
   b. Some measurable amount of resistance
   c. Infinite resistance
   d. OL

8. Signs of overheating at a terminal are often an indication of a(n) _________.

   a. defective fuse
   b. loose connection
   c. overloaded motor
   d. undersized wire

9. Once a system has achieved full-load operation, what is the acceptable voltage drop from the no-load voltage?

   a. 1%
   b. 2%
   c. 5%
   d. 10%

10. If the wiring diagram for the equipment being serviced is missing, what should the technician do?

    a. Order a new service manual from the manufacturer
    b. Replace parts until the unit works
    c. Trace the wires and sketch out a new drawing
    d. Use a wiring diagram from another unit made by the same manufacturer

11. Which of the following is the best method of testing a fuse?

    a. Check the current draw with an ohmmeter
    b. Open the disconnect and check the voltage to the fuse
    c. Remove the fuse and test it with an ohmmeter
    d. Replace the fuse to see if a new one works

12. An infinite resistance reading, from both the run terminal and the start terminal to the common terminal, is most likely an indication of a(n) _________.

    a. defective fuse
    b. defective run winding
    c. defective start winding
    d. open internal overload

13. What type of bearings generally are used in motors designed for heavier-duty operation?

    a. Ball
    b. Needle
    c. Roller
    d. Sleeve

14. The part of the thermostat that contains the function switches is known as the _________.

    a. bimetal
    b. cooling anticipator
    c. heat anticipator
    d. subbase

15. How many magnetic coils does a standard gas valve contain?

    a. One
    b. Two
    c. Three
    d. Four

16. A properly operating door switch on a furnace is very important because it _________.

    a. is usually the first thing to fail
    b. kills power to the whole furnace when it opens
    c. shuts down the fan
    d. turns off the gas valve

ILLUSTRATION CREDITS
FIGURE 6-2: FLUKE CORPORATION
FIGURE 6-10: COOPER BUSSMANN INC.
FIGURE 6-12: HONEYWELL
FIGURE 6-13: ALLANSON ENVIRONMENTAL ELECTRICS
FIGURE 6-18: CAPACITOR INDUSTRIES
FIGURE 6-20: B&K PRECISION
FIGURE 6-21: CARRIER CORPORATION
FIGURE 6-23: CARRIER CORPORATION
FIGURE 6-24: HONEYWELL
FIGURE 6-25: RICH HOKE
FIGURE 6-26: RICH HOKE
FIGURE 6-28: HONEYWELL
FIGURE 6-29: FIELDPIECE INSTRUMENTS
FIGURE 6-30: RICH HOKE
FIGURE 6-31: RICH HOKE

CHAPTER 7

Oil Burners

FUEL OIL

In an oil heating system, fuel oil reaches the burner through a system of tanks, pipes, and pumps. There are six grades of fuel oil, but the most commonly used for residential and light commercial heating is No. 2 fuel oil. No. 2 fuel oil is the accepted product for high- and low-pressure gun burners. It is also used in some vertical rotary vaporizing burners.

Combustion

The goal of the burning process is to generate maximum combustion efficiency. A service technician must be able to adjust the air that mixes with the oil and then measure the products of combustion to determine efficiency. The efficiency measurement for an oil combustion system is a percentage of carbon dioxide (CO_2) in the combustion vent gases. (Check with the burner and system manufacturer for the desired CO_2 percentage.) In order to achieve the optimum CO_2 level, a certain amount of excess air must be introduced to compensate for the portion that does not mix completely with the fuel in the combustion process. For the purpose of heat calculations, remember that one gallon of No. 2 fuel oil equals approximately 140,000 Btu.

Air admitted through the burner shutters mixes with the fuel being forced through the burner tube by the burner blower wheel. In the correct proportions, the air/fuel mixture generally includes two parts hydrogen (H) and one part oxygen (O), thus forming H_2O, or water vapor. If the air/fuel mixture is incorrect, the following can occur:

- The carbon (C) in the fuel that is not completely absorbed in combustion is thrown off as soot.
- Carbon monoxide (CO), a combustible and pollutant gas, may be formed (and released up the venting system) if the carbon present is mixed with oxygen in the proportions of one part carbon and one part oxygen.
- Carbon dioxide may be formed (and released up the venting system) if carbon is mixed with oxygen in the proportions of one part carbon and two parts oxygen—but be aware that the CO_2 created may not be the proper percentage for efficient combustion. It is therefore critical for the service technician to understand the entire fuel oil-burning process.

There must be adequate combustion air in order for combustion to take place. The construction of some new homes may be sufficiently "tight" to require air for combustion to be piped from outside the living space to the burner.

HOW IS OIL IGNITED?

Oil as a liquid is difficult to ignite. But it can be converted to a form that can be ignited easily and produce heat. The procedure most commonly used to prepare oil for combustion is atomization.

Atomization

Atomization is the process of mechanically breaking up oil into small droplets and then mixing these droplets with the air required for combustion. There are two atomization methods in general use—the low-pressure method and the high-pressure method. Both methods are utilized in residential work, but the high-pressure burner is the most common unit found in homes today.

Low-pressure method. In both the high- and low-pressure methods of atomizing oil, oil is broken up by pressurizing it and then passing it through a small orifice or nozzle. The forced ejection through the nozzle atomizes or breaks the oil into small particles and discharges them in such a way as to mix the oil thoroughly with the air required to burn it. The basic difference between high- and low-pressure burners, as the name implies, is the pressure at which the fuel is pushed through the atomizing nozzle. In a

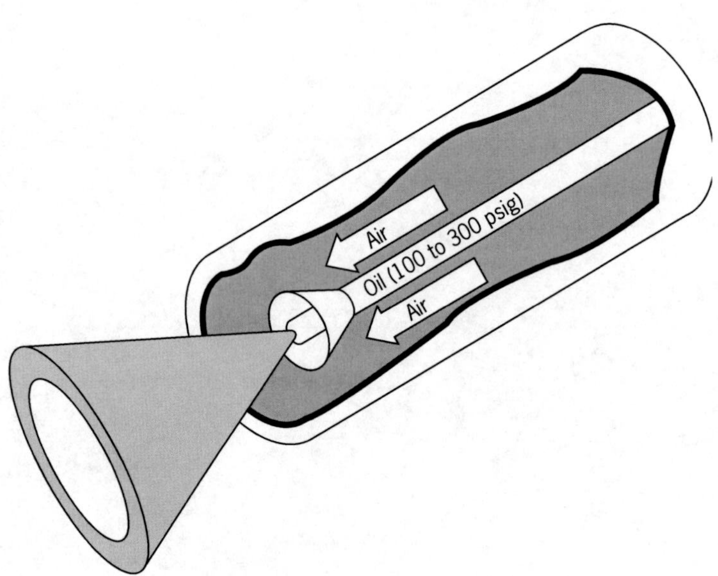

Figure 7-1. High-pressure burner

low-pressure burner, oil and primary air are mixed and forced through the nozzle at a pressure of between 1 and 15 psig. The oil in the mixture is atomized as it passes through the nozzle. The velocity draws (or *aspirates*) the secondary air into the mixture, preparing it for complete combustion.

Operation of the low-pressure burner is satisfactory, but each manufacturer uses a different mechanical method of mixing primary and secondary air with the atomized fuel. This requires special service knowledge and techniques for each different burner unit. Simplicity of service is essential in any residential heating system, so the low-pressure burner is not as common in these applications.

High-pressure method. The high-pressure burner does not use primary and secondary air as such. It is designed to atomize all of the fuel oil and mix it with sufficient air for complete burning in one simple mechanical operation.

The fuel oil is forced through the atomizing nozzle at a pressure of 100 to 300 psig, as shown in Figure 7-1. The velocity of the air as it leaves the retention head creates a low-pressure area (pressure drop) into which the atomized oil will flow. In addition, the combustion air is being pushed by a mechanical fan and the vanes create a turbulence. All of these factors—aspiration by the velocity of oil movement, air pressure, and turbulence—are designed to create complete mixing of the atomized fuel and the combustion air. The mixture of air and atomized fuel is then ignited and leaves the end of the burner as a flame. A high-pressure gun burner is an *assembled* burner. Figure 7-2 shows an example of a modern flame retention burner of a type that is used extensively in residential and small commercial applications.

Figure 7-2. Typical oil burner

VENTING

Technicians who troubleshoot oil burner systems must be familiar with the means used to exhaust the products of combustion to the outside air. The exhaust system consists of the vent connector (or *stack*) and the chimney. During the normal combustion process of a burner, 1 lb of fuel oil is combined with approximately 14 lb of air. In the combustion process, oil and air are converted into an equal weight of flue gases (15.3 lb). The stack and chimney are the exhaust pipe of the system, transporting the flue gases from the boiler or furnace and exhausting (venting) them to the atmosphere.

WHAT IS DRAFT?

Draft is defined by NFPA 31 as "a pressure difference that causes gases or air to flow through a chimney, vent, flue, or appliance." *Natural* draft is "produced by the difference in the weight of a column of flue gases within a chimney or vent system and a corresponding column of air of equal dimension outside the chimney or vent system." *Currential* draft occurs when high winds or air currents blow across the top of a chimney, creating a suction in the stack.

In the majority of oil burner applications, chimney conditions are such that natural draft is adequate for disposing of flue gases. When this cannot be depended upon, *mechanical* draft is produced by installing a motor-driven fan or blower. When the fan or blower *pulls* the flue gases through the chimney or vent, the draft is *induced*. The fan or blower is not of sufficient capacity to handle all of the combustion gases. It serves as a "booster" to augment a draft condition that is otherwise inadequate.

When the fan or blower *pushes* the flue gases through the chimney or vent, the draft is *forced*. A fan or blower of sufficient capacity handles all of the combustion gases. In some instances, it delivers them to a chimney or vent under static pressure. This type of system is seldom found in residential installations in the field. However, it is applied to some models of package units (furnace-burner and boiler-burner assemblies). Keep in mind that insufficient combustion air intake can cause problems with poor draft. If low draft problems exist, be sure that proper combustion air intakes are in place before other measures are attempted.

Why draft is needed

Older pot-type or vaporizing oil burners depend on draft to pull combustion air into the flame. Accordingly, draft is critical with these types of burners, and problems often are due to insufficient draft or poorly controlled draft. Some models with better designs are equipped with a small booster fan, which can eliminate most draft problems.

With newer high-pressure oil burners, combustion air is not dependent on draft. The oil burner has its own motor-driven blower designed to supply the required amount of combustion air to the furnace or boiler.

Measuring draft

The draft created by a vent system is very low. It is measured in inches of water column and is often expressed as a negative value (e.g., an overfire draft reading of –0.01 in. w.c.). Draft can be measured easily with modern, direct-reading gauges. The system manufacturer specifies the draft pressure (positive or negative) for the locations where draft is to be measured. Depending on the manufacturer, those locations are normally "over the fire" and in the flue vent connector (just above the boiler or furnace and before the barometric damper).

A vent system must be warm and dry in order to produce any draft. For this reason, a vent system (whether masonry, metal, or plastic) located on the exterior of a structure may not produce the necessary draft, since it tends to heat up slowly and cool rapidly.

A vent system should extend at least 2 ft above anything within a 10-ft radius, as shown in Figure 7-3. (The authority having jurisdiction, or AHJ, may alter these requirements, so it is best to stay current on local codes.) The reason for the height extension is to prevent an unwanted positive downdraft created by wind currents.

On pitched roofs, the vent system must extend at least 2 ft above any portion of any structure within 10 ft (measured horizontally), and must extend at least 3 ft above the highest point of its roof penetration, as shown in Figure 7-4. (Again, the AHJ may make changes, so follow the current local code.) A vent system should be constructed with minimal turns and be as straight as possible.

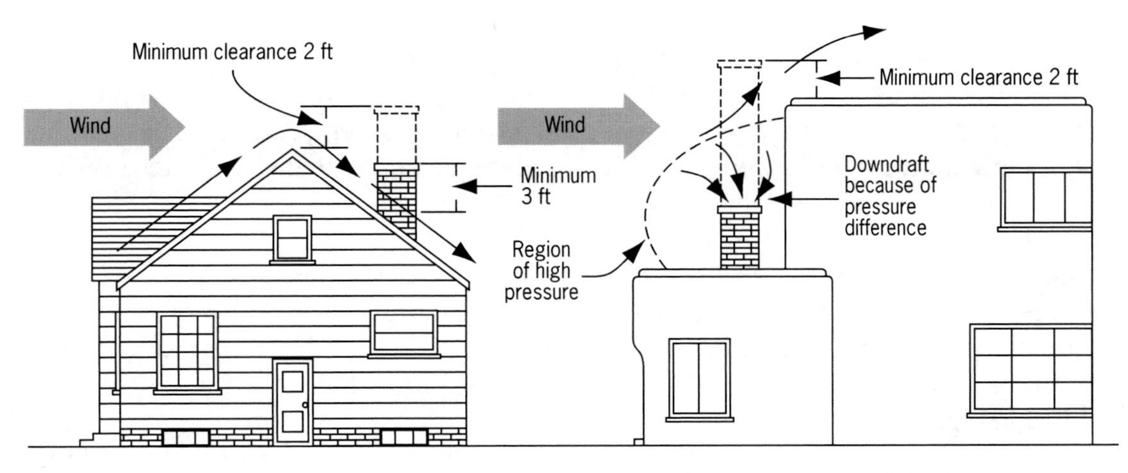

Figure 7-3. Common vent conditions likely to result in backdrafts

Offsets may accumulate dirt, soot, and scale that in time could reduce the cross-sectional area of the vent and impair the draft.

Effect of draft on burner air delivery

The air pressure created by the burner blower, which is not a positive-displacement type, is only about 0.25 in. w.c. in the air tube of the average burner. If the draft in the combustion chamber is –0.10 in. w.c., there will be a pressure of 0.25 in. w.c. pushing air into the chamber, and a vacuum of –0.10 in. w.c. pulling the air in. The total force of 0.35 in. w.c. causes the air to flow.

If the combustion chamber draft drops to –0.01 in. w.c., the total force causing the air to flow becomes 0.26 in. w.c. This is a loss of approximately 25% in the force causing the air to flow. The amount of air flowing into the combustion chamber will be reduced by almost the same amount. Assume that the burner air adjustment was set when draft in the combustion chamber was –0.10 in. w.c. Then the burner would very likely smoke when the draft dropped to –0.01 in. w.c., because of the reduced air flow into the combustion chamber. For this reason, it is necessary to install an efficient draft regulator in the stack to regulate the draft properly. It is important for the draft to be regulated before the burner air adjustment is set.

Because draft does not exist to any great extent during a cold start-up, a burner should not depend on the additional combustion air caused by draft. The best way to be sure that the burner does not depend on this air is to set the burner

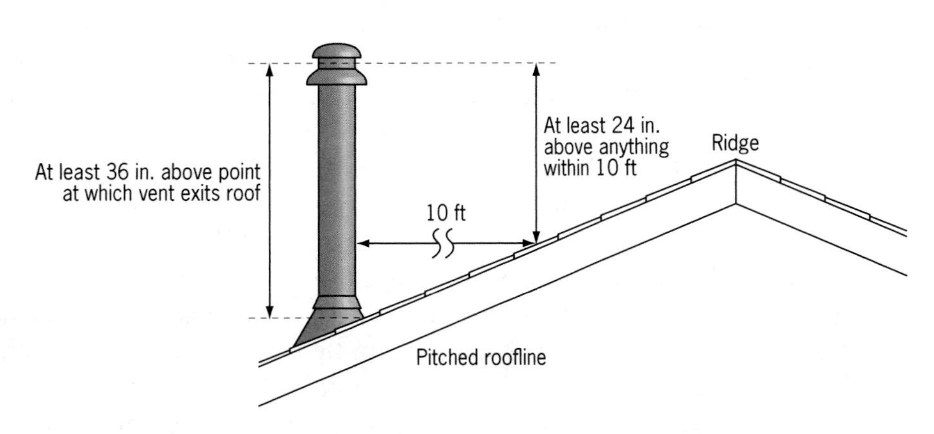

Figure 7-4. Vent system specifications for pitched roof

for zero-smoke combustion with a low overfire draft (from –0.01 to –0.02 in. w.c.). If a quality burner cannot produce a zero-smoke combustion reading under low-draft conditions, there is something wrong with the burner, and it should be corrected. Using a high draft setting to obtain enough combustion air for clean burning is like depending on a crutch that is not always there. A burner that burns clean only with high draft will cause smoke and soot any time the chimney is not producing high draft.

COMBUSTION CHAMBERS

The function of a combustion chamber is to satisfy the requirements of the following "rules of combustion":

- Oil as a fuel must burn completely "in suspension," which means that the flame must never touch any surface. Cold surfaces reduce the combustion temperature and cause soot and carbon formation. No oil should hit the walls of the combustion chamber during the burning process.
- The mixture of air and oil vapor will burn most efficiently in the presence of hot refractory. (*Refractory* is the heat-resistant material used to line the combustion chamber.) Oil must be burned completely, and be converted into hot gases. In order for combustion to be self-sustaining, the heat produced by the flame must be sufficient to ignite fresh mixtures of combustibles (air and oil vapor) that are adjacent to or surrounding the burning zone. The hotter the area around the burning zone (*hot refractory*), the easier the combustibles will burst into flame, and the hotter the flame will be.
- A minimum amount of air must be supplied for complete, efficient combustion. As noted below, this requirement is also affected by the design of the combustion chamber.

The importance of an efficient combustion chamber is dependent on the appliance and cannot be overemphasized. The chamber must meet the manufacturer's specifications and must be:

- made of the right material
- sized properly for the required firing rate
- shaped correctly to match the flame pattern
- the proper height.

If the chamber is too small, if it is of the wrong shape, or if the flame zone is too close to the floor, there will be losses caused by *impingement* (when the flame strikes the sides of the combustion chamber). The combustion chamber can be checked for impingement by visual inspection. If the chamber is too large, or if the flame zone is too far from the floor, the chamber refractory will not become hot enough to reflect the heat back into the flame zone. In addition, more air must be introduced to control smoke, which reduces the temperature of the flame and the CO_2 percentage If the chamber side walls are too low, combustibles will break over the top of the chamber and burn incompletely, because of the lack of hot refractory surrounding them. This causes another loss in efficiency.

You may be called on to evaluate an incorrectly built chamber, or to design and build an efficient one. There may be times when you have to make an exception as to the size and shape because of the physical properties of the heating plant. But otherwise, it should be as close to "perfect" as possible. By this definition, the combustion chamber should, as nearly as possible, be the right size and the right height, with front wall construction placing the nozzle at exactly the right height from the chamber floor. When constructed of the proper material, a combustion chamber should heat up quickly to provide the benefit of a hot refractory early in every burning cycle.

OIL BURNER IGNITION SYSTEMS

The completely automatic electric ignition feature of a high-pressure gun burner is one of the many reasons for its popularity. There are no pilot lights to consume fuel during the OFF cycle, which is an important energy-saving factor. When the ignition system is functioning correctly, flame can be established at any time—whenever the thermostat or other operating control calls for heat.

The term"completely automatic" means that the burner ignition system is unattended. Therefore, its efficient and dependable operation becomes a major safety factor. It is important to have electric ignition the instant oil spray is emitted from the nozzle, as emphasized by the following two examples:

- There may be a weak high-voltage transformer, incorrectly spaced or located electrodes, or dirty or cracked ceramic insulators. If any of these problems exists, the initial oil spray from the nozzle may not be ignited. An excessive volume of volatile vapor could be formed within a few seconds. Exposure of this vapor to a weak spark, or to a satisfactory spark across electrodes that are improperly adjusted with relation to the oil spray pattern, would result in a delayed ignition or an immediate minor explosion.
- In the absence of any electric spark, vapor will continue to form in increased volume until the burner is shut off by the primary control. During the interim period, the vapor could migrate to the stack and into the chimney. If it is then ignited, the result could be potentially hazardous.

It becomes obvious from these scenarios that a thorough knowledge of the components of an automatic electric ignition system and their function is a must if you are to assemble and adjust the various components correctly when installing a burner.

Ignition system assembly

Figure 7-5 on the next page illustrates a typical automatic electric ignition system. Components include the high-voltage ignition transformer, ignition cables or form-connecting bars, ignition electrodes (which are inserted into ceramic insulators), and an electrode bracket fastened to the nozzle pipe (which holds the electrodes in their proper position). A small amount of the atomized oil spray from a high-pressure burner is sufficiently heated by the electric arc, or spark, created by the two properly located electrodes. The high voltage necessary to produce a spark of sufficient heat is supplied by the step-up ignition transformer (which is capable of supplying voltages in the range of 5,000 to 10,000 V).

There are two types of electric ignition systems:

- **Interrupted ignition.** The ignition spark remains on for only a short time at the beginning of each burner operating cycle, and it is turned off by the action of the primary control.
- **Intermittent ignition.** The spark that initially ignited the oil vapors to establish a flame remains on as long as the burner is firing.

The spark across the electrode gap at the tips of the electrodes must be strong enough to withstand the velocity of the air blown through the air tube by the burner fan. Air introduced through the air tube literally forces the ignition spark to form an arc in the direction of the oil spray. This arc should extend toward the spray, causing the oil to be ignited. The result will be an established flame, if the ignition voltage is sufficient to create a spark hot enough to ignite the oil.

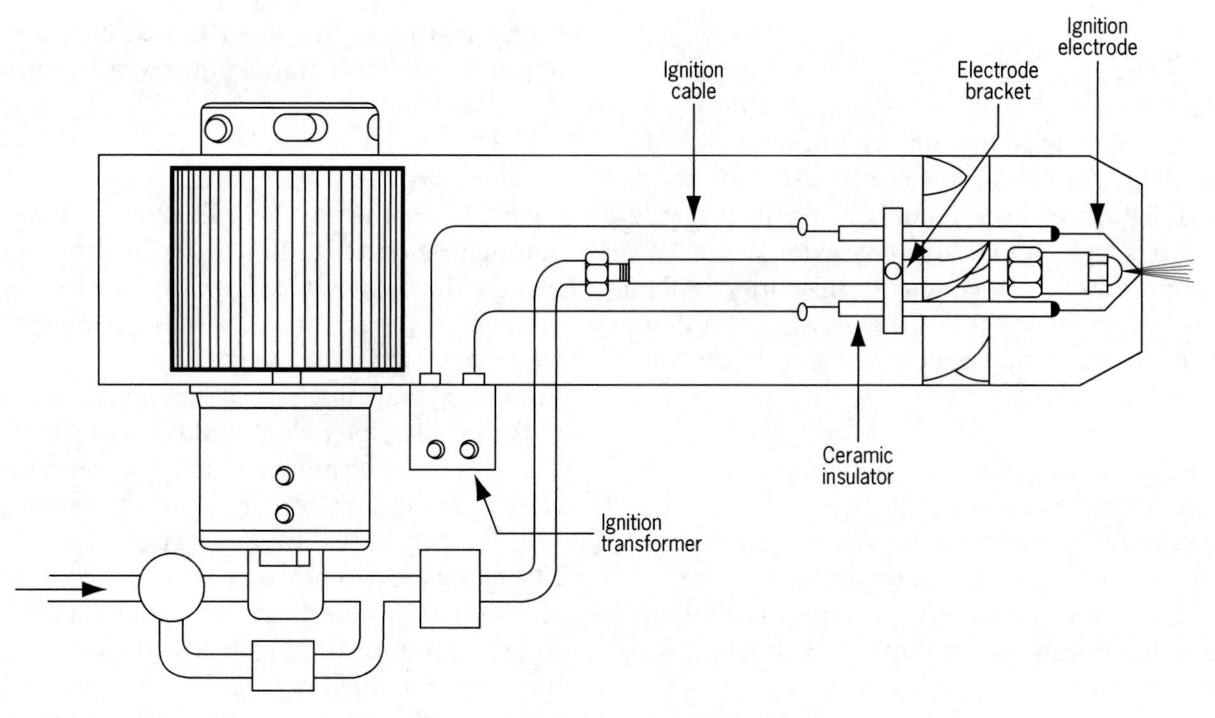

Figure 7-5. Automatic electric ignition system

OIL-BURNING ACCESSORIES

High-pressure fuel pumps

All fuel oil pumps used in high-pressure gun burners are rotary pumps, using gears or cams or a combination of both. The service technician typically is not too interested in the method of compression, because most manufacturers recommend replacement of a defective pump rather than field repair. Detailed knowledge of the pump interior is not required, but there are adjustments that can be made on this pump. Therefore, it is important to understand the general principles upon which this type of pump operates.

A shaft seal is required because these pumps are driven through a shaft from an external power source. For long seal life, the oil pressure on the internal surface of the seal should not be more than 3 psig, according to NFPA 31-8.7.4. Some require the suction pressure on the pump to be 3 psig or less. This is important when a lift pump is used to supply oil to one or more burners.

The purpose of the pump is to create high pressures. The internal surface of the seal is vented to the fuel tank in two-pipe systems. The internal surface of the seal is ported to the suction side of the pump in one-pipe systems.

An automatic pressure valve is an internal part of the fuel oil pump. Its function is to maintain constant oil pressure to the nozzle to ensure efficient operation. The pump actually produces from 100 psi to as much as 300 psi at the discharge.

This pressure-regulating valve, which is usually an integral part of the pump, has an accessible adjusting stem. It also is designed to have positive cut-off as soon as the oil pressure starts to drop for any reason. This keeps oil from leaking out of the nozzle when the burner is shut off, and prevents oil from burning at the nozzle without air, which creates smoke and soot. The soot can collect on the combustion chamber and heat exchanger of the appliance, reducing the appliance's efficiency. If the nozzle drips oil after the burner has shut down, it will also carbonize the nozzle orifice. Figure 7-6 shows a typical high-pressure fuel pump. It may be built as either a single-stage or two-stage pump.

Single-stage pumps. *Single-stage* pumps normally are used when fuel oil is stored above the burner. The fuel oil flows to the burner by gravity. These pumps also can be used for single-pipe installations in which the vacuum is 6 in. Hg or less. You can lift oil 6 ft if the oil pump is horizontally close to the fuel oil tank. The single-stage pump is used to increase oil pressure for use in the burner.

In a single-stage system, there is generally only one pipe from the tank to the burner. The bypass plug located inside the return port must be removed so that oil from the automatic pressure valve or pressure relief can be bypassed into the return side of the pump.

If a single-stage pump is used with a two-pipe system, the pump can operate with a vacuum as low as 12 in. Hg. If the fuel tank is horizontally close to the burner, then a vacuum of 1 in. Hg will lift oil approximately 1 ft. The second pipe carries oil from the return port back to the oil tank. *The bypass plug, if one is required by the manufacturer, must be installed inside the return port so that the oil is not returned to the suction side of the pump.*

Two-stage pumps. *Two-stage* pumps used with two-pipe installations can lift oil as much as 17 ft if the oil tank is horizontally close to the burner. Make sure that the bypass plug (if required by the manufacturer) is installed. The bypass plug is located inside the return port of the oil pump. The first stage of the pump is used to draw the fuel into the pump, and the second stage of the pump is used to supply pressure to the nozzle. A line is connected to the return port of the oil pump and terminates below the oil level in the oil tank. The system is then known as a two-pipe system. Single-stage and two-stage pumps look alike, so make sure that you have the correct pump for the application.

Burner motors

The burner motor not only drives the fuel pump, but also is responsible for driving the fan, which supplies combustion air to the burner. In older systems, the motor speed was 1,725 rpm. In most newer flame retention systems, the motor

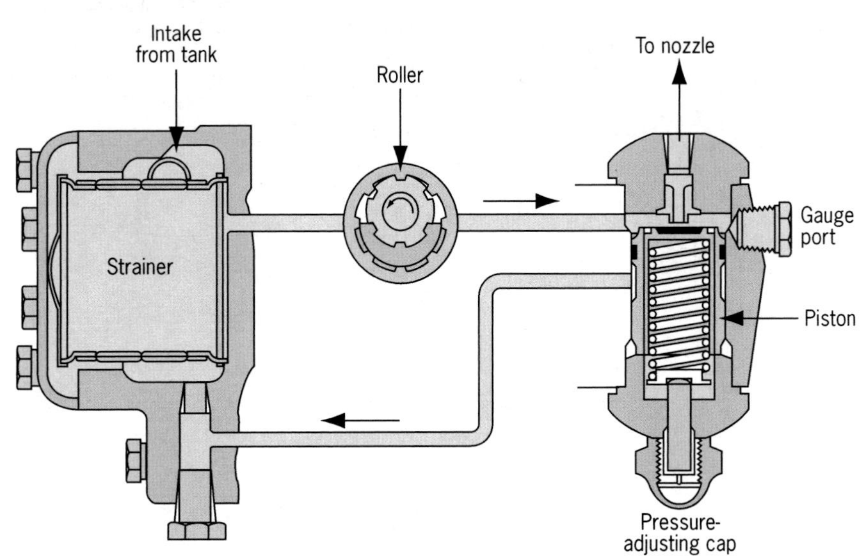

Figure 7-6. High-pressure fuel pump

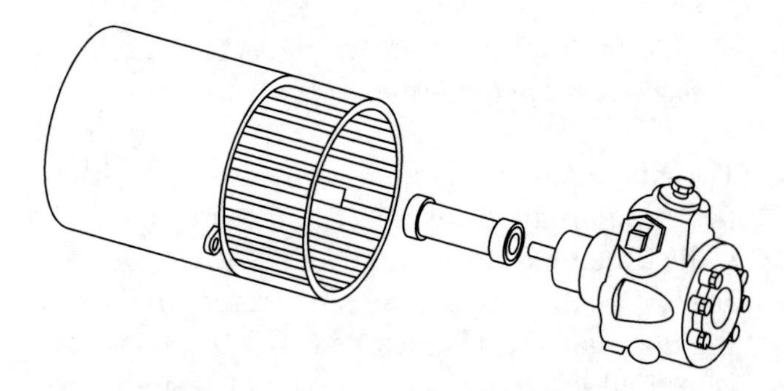

Figure 7-7. Squirrel-cage oil burner fan

speed is 3,450 rpm. Because the fan is turning at twice the speed, greater air pressure can be developed for the flame retention head. The fans of all oil burners of this type are of the squirrel-cage variety, as shown in Figure 7-7. There is little variation in the actual fan wheel itself.

Burners cannot mix air and fuel oil perfectly. They are designed so that the fan will normally provide an excess of air. Because of the variation in nozzle size, fuel oil, etc., an adjustable air intake is provided. Air supply then may be regulated to supply enough air for complete combustion, but not so much as to make the burner inefficient.

Combustion air systems

In addition to the fan and fan motor, the combustion air system of the burner contains a number of very important parts within the air tube. Please refer to the manufacturer's instructions for each of the specialty parts. Figure 7-8 is representative of a residential air tube assembly.

Ignition systems

All high-pressure gun burners use electric ignition, as shown in Figure 7-9. This is accomplished by using a step-up transformer, which supplies high voltage to the electrodes. The ignition transformer shown in Figure 7-10 supplies this high voltage to the pickup contacts of the electrodes. The high voltage creates a spark across the gap between the electrodes. The force of the combustion air in the air tube blows the

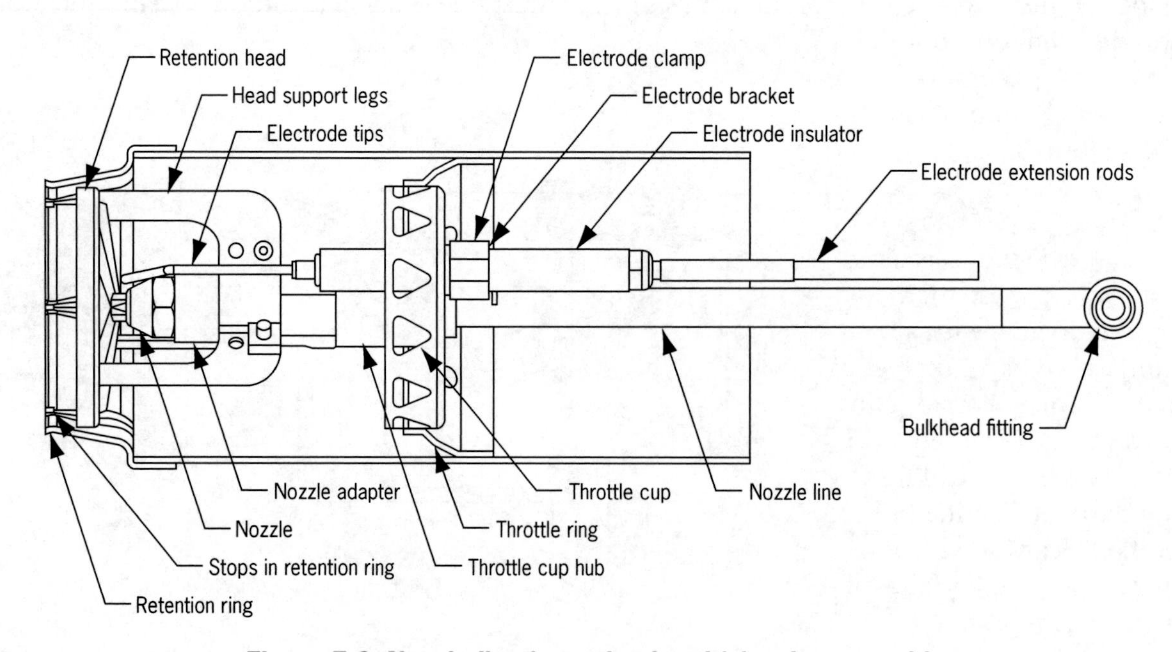

Figure 7-8. Nozzle line/retention head/air tube assembly

spark toward the mixture of oil and air and ignites it. Note that the electrodes are not inserted into the oil/air spray. The spark must be blown toward the spray.

The ceramic insulators shown in Figure 7-9 are designed to insulate the electrodes, but they also serve the purpose of positioning the electrodes. The electrode position is important for ensuring the proper spark arcing. Preferred positioning is covered in the manufacturer's instructions.

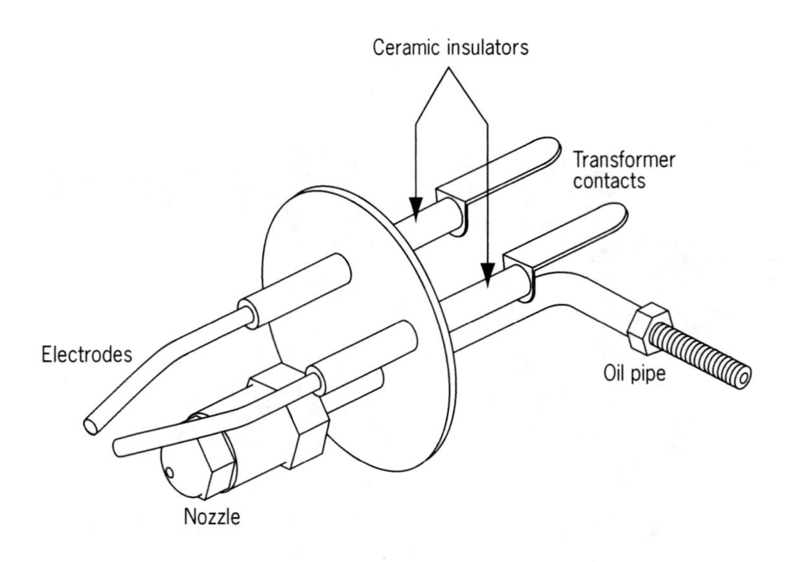

Figure 7-9. Electrode assembly

Iron-core transformers step up a normal 115-V primary to about 10,000 V, and a solid-state igniter can produce 14,000 V. The current developed usually is in the neighborhood of 20 mA (milliamperes). Physical contact with the secondary conductors should be avoided, since nervous reactions to exposure may cause you to make split-second movements and can result in possible injury from contact with adjacent surfaces or objects. The low current means that once the spark has been established, it can be maintained for long periods of time without causing harm or wear on the electrodes.

Ignition can be either intermittent or interrupted. *Intermittent* ignition sparks during the entire burning period. *Interrupted* ignition sparks when the burner is energized, but is shut off after the burner flame is established or after a preset ignition timing period. The ignition method depends on the primary control used to operate the burner. Either option is considered satisfactory.

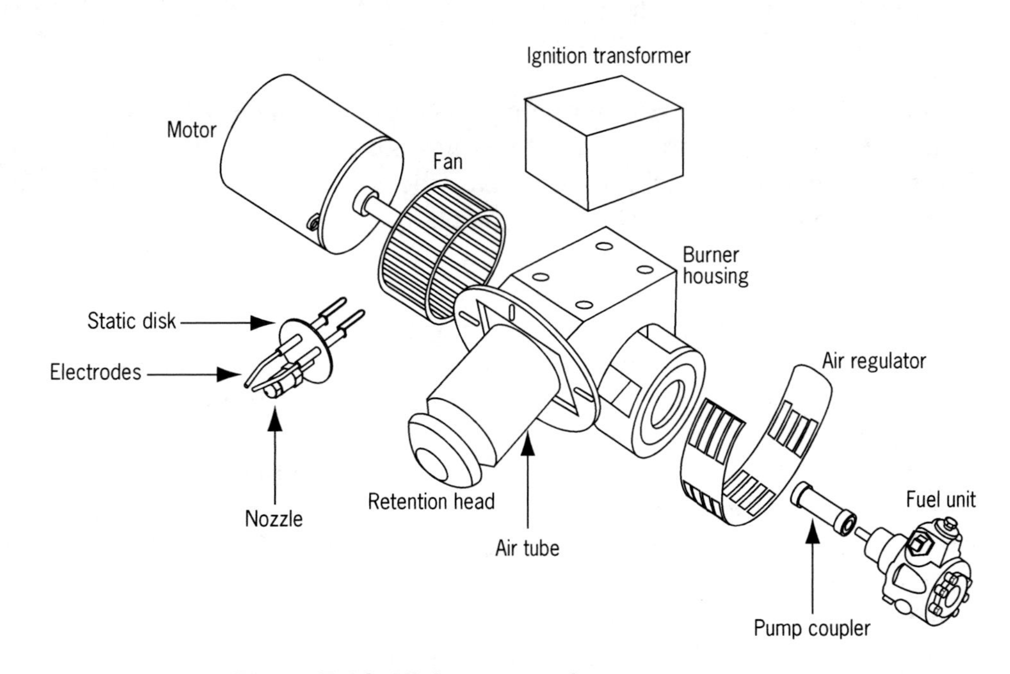

Figure 7-10. High-pressure burner components

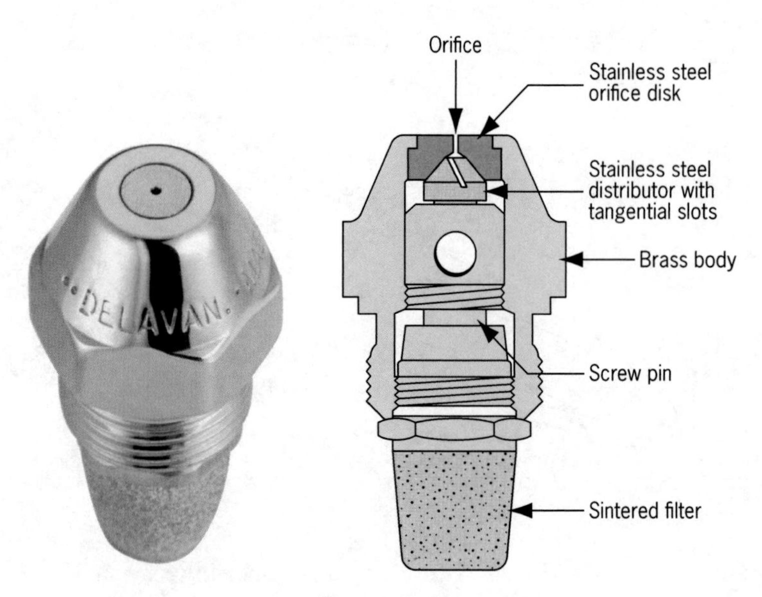

Figure 7-11. Typical oil nozzle

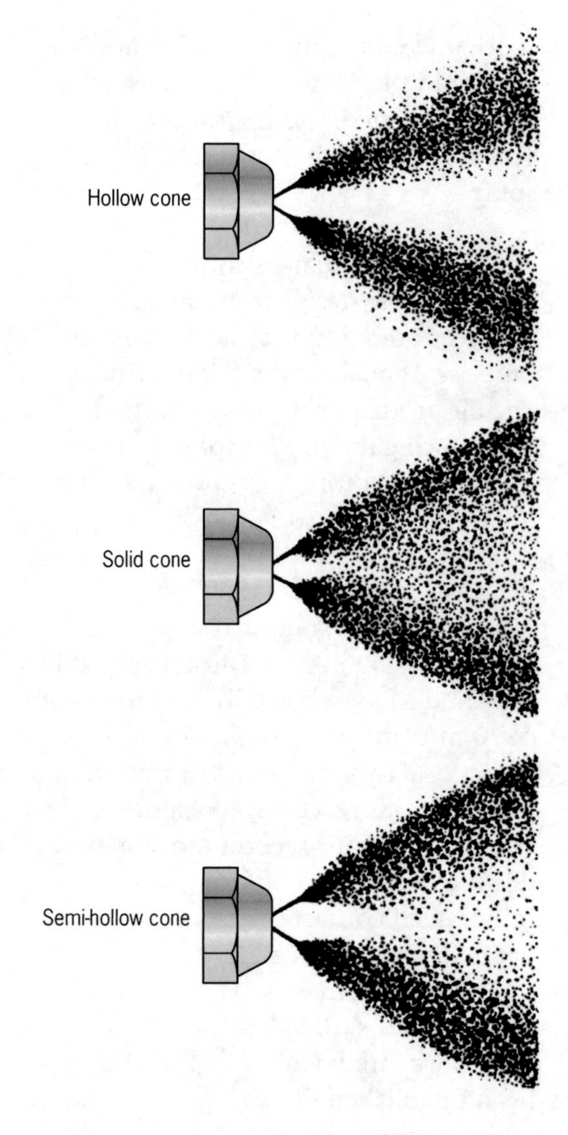

Figure 7-12. Nozzle spray patterns

Fuel nozzles

Once the oil has been brought up to the proper pressure by the pump and has been permitted by the regulating valve to flow into the oil pipe, the nozzle assumes its place of importance. A typical fuel nozzle is shown in Figure 7-11.

In addition to maintaining a fixed flow rate in gallons per hour at the design pump pressure, the nozzle must prepare the oil for mixing with combustion air. It does this by atomizing the oil—or, more correctly, by breaking the oil up into small droplets or a mist, which can be easily mixed with air. When oil enters the nozzle, it first passes through a strainer. Because of the small size of the passages within the nozzle, the smallest speck of dirt can clog or damage it. A fine strainer is a must.

After leaving the strainer, the oil enters the area underneath the tangential slots. As the oil is forced through these slots and into the swirl chamber, a whirling motion is imparted to it. As the oil passes through the nozzle, the decreased area of the nozzle orifice increases its velocity, and the oil breaks up into a fine spray or mist. This mist leaving the nozzle is then mixed with the high-velocity air leaving the air tube choke. Obviously, everything possible is done to ensure a thorough mixing of the oil and combustion air before combustion.

To get the high velocity required for mixing, nozzle orifices are quite small. In the small residential size burner, rates of only 0.5 gal/hr to

2 gal/hr are needed. To meet these requirements, the nozzle must be a finely machined unit built to very close tolerances. A nozzle should be treated as what it is—namely, a piece of precision machinery. Because of the extremely fine tolerances achieved during their manufacture, nozzles are normally replaced. They are rarely serviced in the field unless replacements are not available.

Each application has different requirements for flame types. Therefore, the spray patterns formed by a burner nozzle will vary as illustrated in Figure 7-12. The hollow type is at the top, the semi-hollow at the bottom, and the solid type in the center. The important thing to remember is that the swirling air, the atomization, and the firing chamber all work together to make up a proper system. Always make sure that you follow the equipment manufacturer's instructions regarding nozzles.

When choosing or changing a nozzle, be aware that mismatching of the nozzle and the air choke (of an older burner) or the combustion head (of a flame retention burner) will result in improper firing. If nozzles are replaced with undersized nozzles, the burner may appear to be operating normally but be unable to produce sufficient heat. Oversized nozzles can cause overfiring, which may result in impingement or damage to the heat exchanger. □

REVIEW QUESTIONS

1. What is the most common grade of oil used for residential and light commercial heating applications?

   a. No. 1
   b. No. 2
   c. No. 3
   d. No. 4

2. Which of the following is a product of complete combustion in an oil burner?

   a. Carbon
   b. Carbon monoxide
   c. Carbon dioxide
   d. Carbon trioxide

3. A common procedure used to prepare oil for combustion is called __________.

   a. aeration
   b. atomization
   c. liquidation
   d. optimization

4. A typical oil pressure produced by the oil pump on a high-pressure burner is __________.

   a. 5 to 10 psi
   b. 25 to 50 psi
   c. 50 to 100 psi
   d. 100 to 300 psi

5. In order for oil to burn completely, it must burn __________.

   a. at low pressure
   b. at low temperature
   c. in suspension
   d. with high velocity

6. The combustion chamber of an oil burner must match the __________.

   a. choke nozzle length
   b. draft level
   c. flame pattern
   d. oil pump pressure

7. In which type of oil burner ignition system does the ignition spark remain on as long as the burner is firing?

   a. Direct
   b. Intermittent
   c. Interrupted
   d. Standing pilot

8. If a fuel pump cannot develop proper oil pressure, the most likely cause is a _________.

   a. bad pump seal
   b. dirty oil filter
   c. failed rotary vane
   d. plugged nozzle

9. What type of oil pump system should be used if the oil burner is located *above* the oil tank?

   a. Diaphragm
   b. Single-stage
   c. One-pipe
   d. Two-pipe

10. An ignition transformer of the type used with oil burners can supply a typical secondary voltage of _________.

    a. 1,000 V
    b. 3,000 V
    c. 5,000 V
    d. 10,000 V

11. The primary function of an oil nozzle is to _________.

    a. atomize the fuel
    b. increase the pressure of the fuel
    c. increase the velocity of the fuel
    d. mix air with the fuel

12. When should you clean a high-pressure oil burner nozzle?

    a. At the beginning of each heating season
    b. At the end of each heating season
    c. At least twice a year
    d. Only if a proper replacement is not available

13. Which of the following is an indication that the nozzle on an oil burner is *undersized*?

    a. Cycling on the high limit switch
    b. Impingement of the flame
    c. Insufficient heat output
    d. Overheated heat exchanger

14. Which of the following is an indication that the nozzle on an oil burner is *oversized*?

    a. Clogging of the strainer
    b. Impingement of the flame
    c. Insufficient heat output
    d. Poor atomization

ILLUSTRATION CREDITS
FIGURE 7-2: BECKETT CORPORATION
FIGURE 7-6: CARRIER CORPORATION
FIGURE 7-7: CARRIER CORPORATION
FIGURE 7-8: BECKETT CORPORATION
FIGURE 7-9: CARRIER CORPORATION
FIGURE 7-10: CARRIER CORPORATION
FIGURE 7-11: DELAVAN, INC.
FIGURE 7-12: DELAVAN, INC.

CHAPTER 8

Oil Tanks

INSTALLING OIL TANKS

Although the installation of a storage tank and piping is generally quite simple, it must be done correctly. It must comply with all local codes, regulations, and ordinances. Where these do not exist, the regulations of NFPA Standards 30 and 31 must be observed. All local regulations should be thoroughly investigated before any work is begun. Tank installation must conform with the requirements of the local authority having jurisdiction (AHJ).

There are three possible locations for oil storage tanks:

- inside the basement or furnace room
- outside the building, above ground
- buried underground according to code regulations.

In some communities, the second location (outside the building, above ground) is prohibited except for approved short-term use. An example of short-term use is providing temporary heat in the early stages of construction.

INDOOR TANKS

In residential installations, as well as in some commercial installations, the indoor 275-gal capacity basement tank is the most common. Low cost and long tank life are advantages. Other factors include the ease of installing piping to the burner and the uniform temperature of the product, which ensures constant viscosity of the oil, resulting in cleaner fires.

Figure 8-1 shows a typical tank installation. Note that this illustration is intended for representative purposes only. Two-pipe systems normally are used with exterior tank locations.

Generally speaking, tanks are manufactured in 275-gal and 330-gal sizes for residential installation. The more common 275-gal tanks are available in two configurations, referred to simply as "horizontal" and "vertical." The horizontal tank is approximately 27 in. high × 44 in. wide, and the vertical tank has the inverse dimensions (see Figure 8-2).

Both sizes are made so that they can be carried through doorways, etc. A tank must be located at least 5 ft away from any open flame or fuel-burning appliance, and at least 3 ft away from any panel box or meter. In some communities, this distance requirement is extended to 7 ft. The height of the tank off the floor can vary. Follow the manufacturer's instructions (which typically specify that the legs of the tank support can be

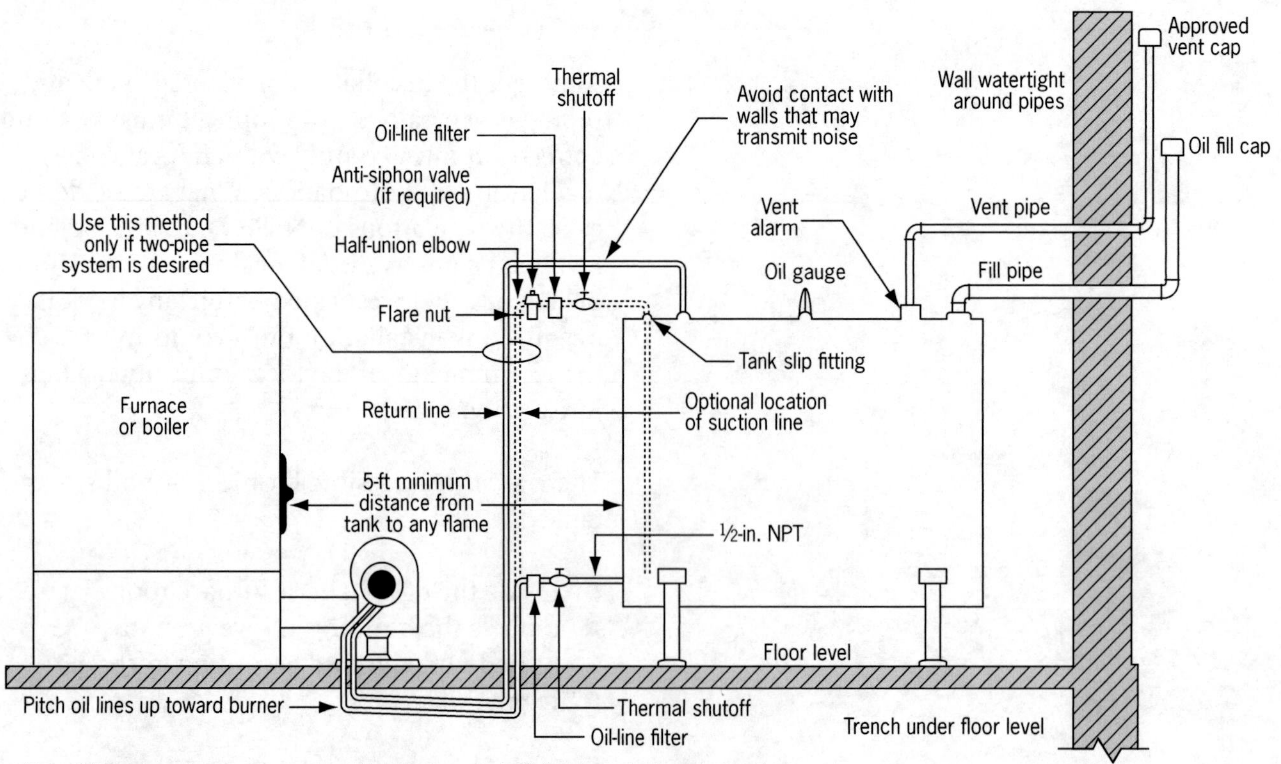

Figure 8-1. Typical tank installation

no longer than 12 in.). The legs are screwed into flanges on the bottom of the tank. Floor flanges with bolt holes in them should be installed on the bottom of the legs, so that the tank can be anchored to the floor. The tank shape is *obround*, which makes passage through doorways easier. The tank should be mounted on a noncombustible material, such as concrete or a brick basement floor. Tanks may be installed on other types of floors where AHJ requirements and NFPA codes allow.

If the jurisdiction agency requires the installation to follow NFPA rules, and if the storage tank can be filled so that the oil level is 6 ft or more above the burner, it may be necessary to install a pressure-reducing valve or an oil safety valve (see Figure 8-3). NFPA regulations state that a means must be provided to limit the oil pressure at the burner inlet to a maximum of 3 psi. A suitable UL-approved gauge should be installed in the top of the tank to indicate the oil level at all times (see Figure 8-4 on the next page).

Figure 8-2. Common 275-gal vertical fuel tank

The fill line for the tank should be installed according to NFPA requirements—i.e., the fill pipe shall be located to permit easy filling

Figure 8-3. Oil safety valves

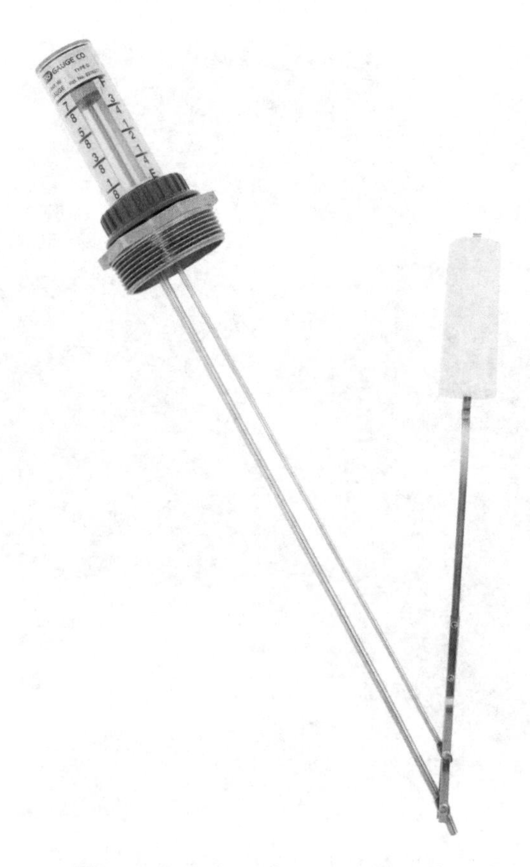

Figure 8-4. Level gauge for oil tank

without spillage, it shall be no less than 2 in. in diameter, it shall be pitched toward the tank, it shall not have sags or traps, and it shall be protected from damage. In addition, the fill line must terminate at least 2 ft from any building opening. The end of the fill line must be equipped with a tight metal cover that resists entry of foreign substances (see Figure 8-5) and must be clearly marked as the heating fuel fill piping.

Figure 8-5. Cast iron fill cap

Vent piping shall be large enough to permit proper filling of the tank and emergency venting if necessary, but no less that 1 1/4 in. nominal pipe size. The vent termination must be within sight of and no more than 12 ft away from the fill piping. The vent pipe must be sized according to the NFPA capacity chart, pitched toward the tank, protected from damage without sags or traps, and without obstructions except for the audible vent alarm.

Further, the vent pipe must terminate outside the building at least 2 ft from any opening. It must terminate high enough to prevent blockage from snow and ice, and must be equipped with a corrosion-resistant weatherproof cap that has a free open area equal to the cross-sectional area of the vent pipe and a mesh screen as required by NFPA regulations.

Figure 8-1 shows the various aspects of fill and vent pipe installation. Before the vent pipe is installed, a vent signal or whistle shall be located per the manufacturer's instructions. When oil is introduced into the tank, air expelled through the vent signal causes it to emit a whistling sound until all the air in the tank is vented, indicating that the tank is full of oil.

As shown in Figure 8-1, a globe or ball valve should be installed at the tank outlet from which the oil feed line (also known as the *suction line*) is run to the burner. On older tanks, the outlet tapping is located in the tank end plate, about 2 in. above the tank bottom. Tanks with an opening in the actual tank bottom are now standard. The older type, with the end outlet, permits an accumulation of water and sludge to build up in the tank bottom, beneath the oil, which shortens the life of the tank.

Figure 8-6. Oil-line filter

NFPA code regulations require a tank to be pitched toward the outlet at a slope of 1/4 in. per foot. A tank installed in this manner allows small droplets of water caused by condensation to reach the bottom of the tank, where they can pass on to the burner and be consumed with no difficulty. A cartridge filter similar to the one shown in Figure 8-6 must be installed in the suction line, at the burner side of the fusible-link safety shutoff.

For the average residential high-pressure gun burner installation, 3/8-in. or 1/2-in. OD copper tubing is adequate, depending on the amount of oil to be transported and the distance, if any, that the oil must be lifted. The oil line has a valve and a filter between the tank and the oil pump. Be sure to replace the oil filter during each annual maintenance. It is important to purge the air out of the line after changing the filter to re-prime the oil pump.

Do *not* use compression fittings and Teflon® tape for installing fuel oil lines. Compression fittings do not seal tightly enough and may let air into the system. Teflon tape can be dissolved by fuel oil. The Firomatic™ valve like one of those shown in Figure 8-7 represents a safety factor in that it is spring-closed. This valve should be mounted close to the inlet of the fuel pump but outside the furnace or appliance. It is held open only by a fitted sleeve of low-temperature metal in the handle, which will

Figure 8-7. Firomatic fusible valves minimize fire hazards

melt at 160 °F. When the low-temperature metal melts, the valve will close, cutting off the oil supply from the fuel oil tank. If a two-pipe system is to be used, there are internal changes that must be made to the fuel unit or pump.

OUTDOOR TANKS ABOVE GROUND

An outdoor tank above ground may be approved under certain conditions. This type of location may be more economical in the case of homes built on slab foundations or with mobile homes. When installing such a tank, it is important to provide a substantial masonry foundation on which to set the tank legs to ensure stability. The means to contain fuel oil spills may be required. On larger installations, retain a competent environmental firm to make sure that you follow all of the rules and regulations.

An outdoor tank installed above ground can be a source of several operational problems. In cold climates, the tank is exposed to changeable weather elements and responds to changes in temperature. For example, any moisture in such a tank can cause frozen oil lines from the tank to the burner. The accumulation of water in the tank may be considerable, because of wide variations of temperature inside the tank. As the tank cools, the oil and the air above it contract. This can suck in moisture-laden air through the tank vent. Water vapor condenses on the inside surfaces, flows down, and settles at the bottom of the tank, below the oil (because water is heavier than oil). On a warm day, the tank and its contents increase in temperature, and air in the tank is expelled. This makes room for a fresh supply of moist air to replace it when the temperature drops again. This type of repetitive cycle can be a genuine source of trouble. If such a tank is used, the line filter in the suction line to the burner must not be installed outside the building, but inside near the burner.

UNDERGROUND TANKS

For cases in which initial installation cost is not a factor, and floor space is at a premium, installation of an underground tank is possible. The installation of an underground tank is a major undertaking. Regulations relating to underground tank installation vary greatly from jurisdiction to jurisdiction. Unless you have a comprehensive knowledge of all of the rules and regulations of the Environmental Protection Agency (EPA) and the local authority in regard to installing underground tanks, you must retain an environmental company that is familiar with all of the relevant requirements.

According to the EPA (40 CFR 280.20 Ch. I), each tank must be properly designed and constructed, and any portion underground that routinely contains product must be protected from corrosion. A major concern of underground tanks is that a leak in the tank or piping could easily go unnoticed for an extended period of time. An oil leak could contaminate the soil around the tank and eventually the ground water. Many jurisdictions encourage the removal of existing residential oil tanks and some even prohibit the installation of new underground tanks. □

REVIEW QUESTIONS

1. All oil tanks must carry the _________ label.

    a. DOT
    b. EPA
    c. NFPA
    d. UL

2. What is the capacity of a typical above-ground indoor oil storage tank?

   a. 150 gal
   b. 275 gal
   c. 550 gal
   d. 750 gal

3. If the oil level in the tank is more than 6 ft above the oil burner, what additional component may be required?

   a. Pressure cutout valve
   b. Pressure differential valve
   c. Pressure-reducing valve
   d. Pressure-regulating valve

4. The vent pipe for an oil tank should have a *minimum* diameter of __________

   a. 1 in.
   b. 1¼ in.
   c. 1½ in.
   d. 2 in.

5. The vent whistle on an oil tank lets you know that __________.

   a. a high-pressure condition exists in the tank
   b. filling is complete and the tank is full
   c. oil is escaping from the tank
   d. the oil level in the tank is low

6. How frequently should the oil filter in the oil line to the burner be replaced?

   a. At least annually
   b. Every time the oil tank is filled
   c. Whenever the pressure drop exceeds 5 psi
   d. Whenever the pressure drop exceeds 10 psi

7. What type of fitting typically is used for copper oil lines?

   a. Compression
   b. Flared
   c. Swedged
   d. Threaded

8. What is the main disadvantage of an underground tank?

   a. Air may be pulled in and out of the tank with changes in temperature
   b. Leaks in the tank may go undetected
   c. Moisture may condense and freeze in the oil lines
   d. Oil temperatures may be inconsistent

9. Which agency generally has jurisdiction over the installation of underground tanks?

   a. DOT
   b. DOE
   c. EPA
   d. UL

10. Underground tanks require __________.

    a. containment vaults
    b. corrosion protection
    c. fiberglass coating
    d. oversized vent lines

ILLUSTRATION CREDITS

FIGURE 8-1: NATIONAL OIL FUEL INSTITUTE
FIGURE 8-2: LIQUIDYNAMICS
FIGURE 8-3: (LEFT) WEBSTER FUEL PUMPS & VALVES
(RIGHT) SUNTEC INDUSTRIES INC.
FIGURE 8-4: KRUEGER SENTRY GAUGE
FIGURE 8-5: LIQUIDYNAMICS
FIGURE 8-6: GENERAL FILTERS, INC.
FIGURE 8-7: HIGHFIELD MANUFACTURING COMPANY

CHAPTER 9

Planned Maintenance

BURNER OPERATION

Let's take a look at the sequence of operation for a typical oil furnace. On a call for heat, the burner motor starts, which brings on both the oil pump and the draft fan. The pre-purge cycle, if required by the manufacturer, allows time for air to begin flowing through the combustion chamber to establish draft, clear out any flammable vapors, build oil pressure, and establish spark for ignition. (*Note:* Not all burners have a pre-purge cycle built into the control system. Some older models may have a time-delay oil solenoid, which accomplishes the same result.)

A dark combustion chamber means that the cad cell has a high resistance and oil begins to flow to the burner. Oil will continue to flow until the control system fail-safe timer locks out the system (if flame is not established). If the cad cell "sees" (verifies) a flame, the cad cell resistance drops and the system continues to operate until the call for heat has been satisfied.

PROPER OIL FLAME

A uniform oil spray pattern produces the proper flame shape. The flame should burn relatively quietly, with no oil or flame impingement on the walls of the combustion chamber. A flame that burns with continuous yellow tips generally is not getting enough combustion air. Carbon buildup on the walls of the combustion chamber may be an indication that the oil pressure is too high. Carbon buildup also may be a sign that the size or angle of the nozzle is incorrect, or that the firing rate needs to be adjusted. Flame impingement on the combustion chamber can be checked visually, because it can cause a burner to smoke after shutdown.

OIL PUMP PRESSURE CHECK

After you have checked that the nozzle is sized properly for the unit, the next step is to check the oil pump pressure. Consult the specifications on the unit data plate or in the manufacturer's literature. Some manufacturers still use 100 psi, but others use different pressures.

COMBUSTION EFFICIENCY TESTING

When a newly installed oil burner is started and the various operating controls and fuel-transporting systems checked, the lapsed time is usually sufficient for the flue passages, smoke pipe or stack, and chimney to obtain a

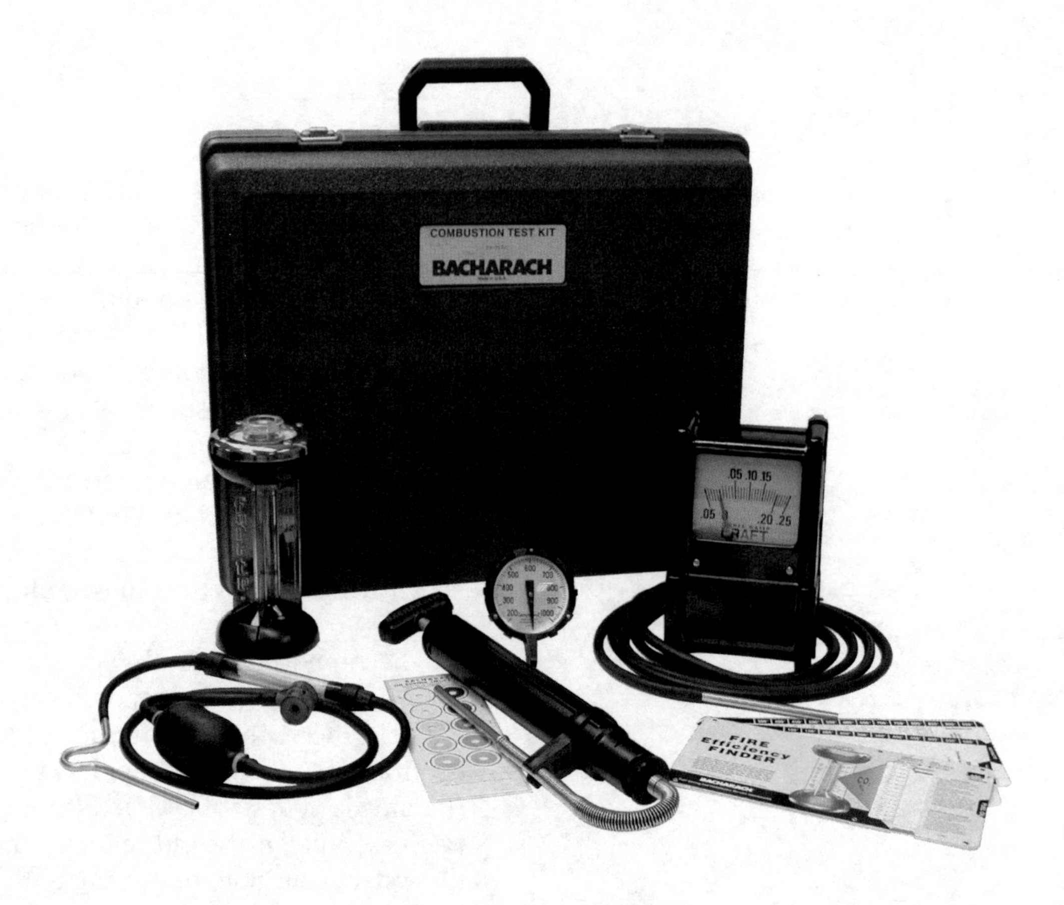

Figure 9-1. Typical oil burner combustion test kit

temperature at which conclusive combustion efficiency tests can be made (called a "steady-state" temperature).

Performing these tests is one of the most important responsibilities of the installer. Such tests cannot be made without the use of appropriate testing instruments and gauges. These testing tools include a smoke tester, stack thermometer, draft gauge, and CO_2 or oxygen indicator. Also required are appropriate combustion efficiency charts and/or a special slide rule calculator with which to combine the results of the various required tests and determine combustion efficiency or the lack of it. Examples of required combustion testing equipment are shown in Figure 9-1 (and later in this chapter in Figures 9-5 and 9-6).

The factors involved in achieving the full benefit of combustion testing are:

- Use a complete combustion testing kit.
- Use a carefully planned, systematic, combustion testing procedure.
- Use systematic, routine methods of recording test data.

Always use a complete combustion test kit. No single instrument will compensate for the lack of other instruments.

COMBUSTION TESTING INSTRUMENTS

Combustion testing instruments must be used in sequence. The first test and adjustment to be made is the draft test and adjustment. Follow the manufacturer's installation recommendations when adjusting the over-the-fire draft of residential and light commercial oil burners. CO_2 readings generally should be around 12%, with a preferred smoke reading of 0 (zero). Again, follow the manufacturer's instructions.

Measure the temperature of the flue or stack after all of the draft and burner adjustments have been made. You can use flue gas thermometers like those shown below in Figure 9-2. Handheld digital instruments are also available that can measure stack temperatures. Remember that the *net stack temperature* can be calculated by subtracting the room temperature from the measured stack temperature. Consult the manufacturer's specifications to obtain the normal net stack temperature for a given oil furnace.

Stack thermometers

The *stack thermometer*, in conjunction with other required combustion testing instruments, can be used to evaluate and solve combustion process problems. The effects of a stack temperature that is too high or too low on the overall combustion efficiency of an oil burner installation—and the possible causes of such a condition—are shown in Table 9-1 on pages 114 and 115.

Figure 9-2. Flue gas thermometers

Overfire draft	
Condition 1: Overfire draft is too low (less than –0.01 in. w.c.). **Result:** 1. Odors. 2. Pulsations, poor ignition.	**Caused by:** 1. If stack draft also is low (–0.02 to –0.04 in. w.c. or less): a. Defective or blocked chimney. b. Leaks in smoke pipe. c. Defective or improperly adjusted draft regulator. 2. If stack draft is high (–0.05 in. w.c. or more): a. Soot in heat exchanger. Unit needs cleaning. b. Poorly designed unit. Might need even higher stack draft.
Condition 2: Overfire draft is too high (greater than –0.02 in. w.c.). **Result:** 1. High heat loss when unit is not firing. 2. Gas flow too high through furnace (poor efficiency).	**Caused by:** 1. Defective or improperly adjusted draft regulator. 2. No draft regulator present. 3. Draft inducer being used when not required.
CO_2 in flue gas	
Condition 1: CO_2 is too low (less than 8%). **Result:** 1. Heat loss up chimney (poor efficiency).	**Caused by:** 1. With low smoke: a. Underfiring combustion chamber. b. Nozzle is too small. c. Air leaks into furnace. d. Air gate open too wide. 2. With high smoke: a. Faulty nozzle operation. b. Poor combustion chamber.
Condition 2: CO_2 is too high. **Result:** 1. Smoke may be too high (greater than "zero" smoke test reading). 2. Pulsations and other noises may occur, particularly on cold starts.	**Caused by:** 1. Not enough air (air gate closed too far). 2. Dirty fan or air-handling parts. 3. Not enough draft. 4. Burner being overfired.

Table 9-1. Oil burner operation diagnosis chart

Smoke in flue gas	
Condition: Smoke too high (greater than "zero" smoke test reading). **Result:** 1. Soot deposits in furnace (causes high stack temperatures and poor efficiency). 2. Soot and smoke discharge from home chimney.	**Caused by:** 1. If CO_2 is high (greater than 12%): a. Overfiring of unit (oil rate too high). b. Not enough excess air. c. Dirty fan or air-handling parts. 2. If CO_2 is low (less than 8%): a. Faulty nozzle operation. b. Combustion chamber trouble (too large, poorly installed, broken, etc.).
Flue gas temperature	
Condition 1: Stack temperature too high (greater than 630°F net). **Result:** 1. High heat loss up chimney (excessive oil consumption).	**Caused by:** 1. Dirty heat exchanger (possibly high smoke and high draft loss). 2. Overfiring furnace (accompanied by high CO_2). 3. Poor furnace design (baffles needed). 4. Poor combustion chamber. 5. Excessive draft.
Condition 2: Stack temperature too low (below manufacturer's specifications). **Result:** 1. Moisture condensation (rusting and deterioration of smoke pipe and chimney). 2. Poor draft.	**Caused by:** 1. Underfiring furnace.

Table 9-1. Oil burner operation diagnosis chart (continued)

The procedure for reading flue gas temperatures in the stack is as follows:

1. Drill a 1/4-in. hole in the flue pipe about 12 in. from the boiler breaching on the furnace side of the draft regulator and at least 6 in. away from the regulator.
2. Turn the burner on and allow it to run until it reaches steady-state operation (this usually takes 5 to 10 minutes) before beginning the tests.
3. Insert the thermometer stem into the test hole. If it is the clip-holding type, use the clip to secure it to the flue.
4. Read the temperature on the dial scale at which the pointer finally comes to rest.
5. Determine the net stack temperature by subtracting the ambient air temperature from the thermometer reading.

High readings are cause for special concern and adjustments. Consult the specification sheet

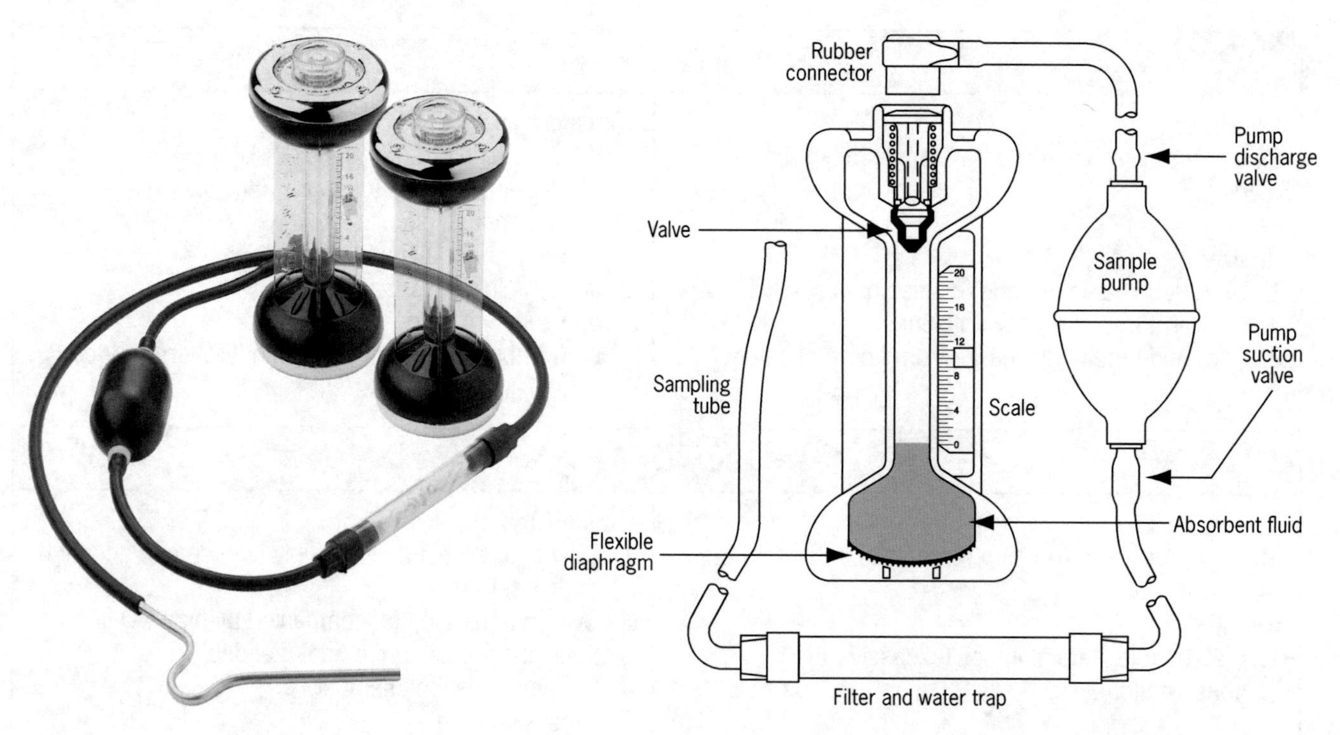

Figure 9-3. CO_2 indicator

Figure 9-4. Construction of gas analyzer

from the manufacturer for individual units. A high stack temperature may indicate any of the following conditions:

- excessive draft through the boiler
- dirty, carbon-covered boiler or furnace heating surfaces
- lack of sufficient baffling in flue passages
- undersized furnace
- incorrect or defective combustion chamber
- boiler or furnace overfired
- improper adjustment of the draft regulator.

Immediately check and remedy these conditions to maintain your goal of peak efficiency.

CO_2 indicators

A key factor in combustion efficiency is the percentage of CO_2 in the flue gas. An example of a CO_2 indicator is shown in Figure 9-3. This instrument is used to sample the combustion gases and determine the CO_2 percentage. (Notice that the testing fluid in the "dumbbells" is *red* for checking CO_2, and *blue* for checking oxygen.) If the CO_2 reading is low, it may be an indication that there is too much excess air, or that the fuel is not burning completely. Adjustments are required. The construction of a gas analyzer is illustrated in Figure 9-4. Electronic instruments, such as the one shown in Figure 9-5 and the handheld model pictured in Figure 9-6, are also available for measuring CO_2.

Although the CO_2 indicator should be used in conjunction with all of the other testing devices, a special *slide rule calculator* correlates stack temperatures and CO_2 percentages. It is used to determine combustion efficiency and stack loss in an oil heating installation. If you use an electronic combustion analyzer to calculate the combustion efficiency, follow the instrument manufacturer's instructions.

A low CO_2 reading is indicative of one or more of the following conditions that must be remedied:

- high draft
- excess combustion air
- incorrect or defective fire box
- air leakage
- poor oil atomization
- worn, plugged, or incorrect nozzles
- excessive air leaks in the furnace or boiler
- incorrect air-handling parts
- erratically operating draft regulator
- incorrectly set oil pressure.

If there is too great a difference in the CO_2 readings between the samples taken in the flue pipe and through the fire door, an air leakage or other unsatisfactory combustion condition exists within the boiler or furnace.

Two factors determine heat loss in the flue gases—the percentage of CO_2 in the gases and the stack temperature. The loss of heat in the flue gases, in turn, determines the combustion efficiency of the oil heating plant. A combustion efficiency chart or slide rule provides a rapid, simple means of determining combustion efficiency and stack loss from the results of the CO_2 and stack temperature tests.

The slide rule has horizontal and vertical slide inserts. You can see a picture of a fire efficiency finder/stack loss slide rule in Figure 9-1. Move the horizontal slide until the stack temperature previously determined appears in the window marked "stack temperature." Next, move the vertical slide until the black arrow points to the CO_2 percentage previously determined. Percent combustion efficiency and stack loss are then

Figure 9-5. Combustion analyzer

Figure 9-6. Residential combustion analyzer

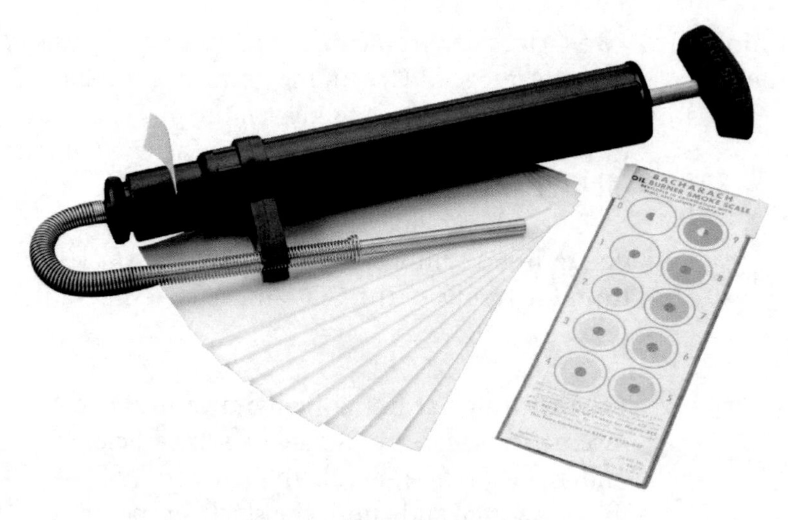

Figure 9-7. Smoke spot tester

indicated in the cut-out of the arrow on the vertical slide. Of course, an electronic testing instrument will perform the efficiency calculation internally and provide a digital readout (see Figure 9-6).

Smoke spot testers

The *smoke spot tester*, shown in Figure 9-7, is used to provide the service technician with an accurate indication of the smoke content in flue gases. Excessively smoky combustion has long been recognized as an indication of wasteful, incomplete, and inefficient oil burner operation. A smoky fire results in excessive soot formation, which is, in effect, an insulator that impedes the transfer of heat from the flame to the water or air in the boiler or furnace. A soot buildup on the heating surfaces is not only wasteful ($^1/_8$ in. of soot can reduce heat absorption by as much as 10%), but also can be the cause of many service difficulties.

The objective of the smoke test procedure is to measure the smoke content in the flue gas and then, in conjunction with other combustion test results, to adjust the burner to optimum operation. The smoke scale used in conjunction with the spot tester has ten color-graded spots. These spots range from 0 (zero), which is pure white, to 9, which is the darkest color. A widely used oil burner smoke scale is pictured with the tester in Figure 9-7.

Table 9-2 shows the potential for soot formation in various levels of smoke as indicated by smoke scale numbers. It should be emphasized, however, that not all oil burners will be affected equally by the same smoke content in the flue gas. Depending on the construction of the heat exchanger or the boiler, some units will accumulate soot rapidly at smoke No. 3, while the accumulation of soot on other units at the same smoke scale reading may be relatively slow. The smoke reading on a flame retention burner should be 0. The following list identifies some probable causes of excessive smoke content in flue gases:

- improper fan delivery
- insufficient draft
- poor fuel supply
- oil pump not functioning properly
- defective nozzle or nozzle of incorrect type
- excessive air leaks in the boiler or furnace
- improper fuel-air ratios
- defective fire box
- improperly adjusted draft regulator
- improper burner air-handling parts.

Draft gauges

Correct draft is essential for efficient burner operation. While draft is not directly related to combustion efficiency, it does affect oil burner efficiency. Draft requirements vary with each type of installation. Consider the following in regard to draft levels:

- Excessive draft can increase the stack temperature and reduce the percentage of CO_2 in the flue gases.
- Insufficient draft may cause pressure in the combustion chamber, leading to the escape of smoke and odor to the basement area.
- Insufficient draft also makes it impossible to adjust the burner for maximum efficiency, because this level of efficiency depends on the proper amount of air mixing with the correct amount of oil—no more and no less—each time the burner runs.

A *draft gauge* (see Figure 9-8) provides direct readings in 0.01-in. w.c. increments. Negative draft readings range from 0 to –0.25 in. w.c., and positive draft readings from 0 to 0.05 in. w.c. Draft gauges can be used to check both overfire draft and flue or breech draft. Some slope gauges (or inclined manometers) can read draft conditions as low as 0.005 in. w.c. The advantage of this type of instrument is that it does not move as quickly with rapid changes in draft. In other words, the reading is dampened, which gives an average reading. This can be an advantage when adjusting draft for an oil burner.

When checking overfire or flue draft, always pay attention to the equipment manufacturer's requirements, since they can vary depending on the equipment and installation. Following the instrument manufacturer's instructions for using a draft gauge is equally important.

The reason for testing for flue pipe draft at a location between the draft adjuster and the boiler breaching is that the position of the counterweighted damper in the draft adjuster is the medium for increasing or decreasing the flue pipe draft level. Adjusting this damper toward

Smoke scale spot No.	Rating	Sooting produced
0	Excellent	Extremely light if at all
1–2	Fair to poor	Slight sooting which will not increase stack temperature appreciably
3	Poor	May be some sooting, but will rarely require cleaning more than once a year
4	Poor	Borderline condition—some units will require cleaning more than once a year
5	Very poor	Soot rapidly and heavily

Table 9-2. Effects of smoke on burner performance

Figure 9-8. Draft gauge

the closed position increases flue pipe draft, and adjusting it toward the open position decreases flue pipe draft.

The importance of utilizing an efficient draft regulator cannot be overemphasized. Too many service technicians assume that all draft regulators will do about the same job. This assumption is false. Make sure that the draft regulator being used is a reliable, highly engineered component, capable of producing a reasonably constant overfire draft for a wide range of chimney draft conditions.

INTERPRETING COMBUSTION EFFICIENCY TEST RESULTS

A relatively high level of combustion efficiency can be achieved by carefully following the system and equipment manufacturer instructions. Failure to follow such instructions will result in a need for corrective measures. Consult the diagnostic charts in Table 9-1 on pages 114 and 115 for probable conditions and causes. The charts include the following factors:

- overfire draft
- smoke in the flue gas
- CO_2 in the flue gas
- flue gas temperature.

In the final analysis, the competence of an oil burner installer will be judged by the long-term dependability and cost-effective operation of the system. Strict attention to final tests and adjustments is necessary, as noted in the various factors covered in the charts of Table 9-1.

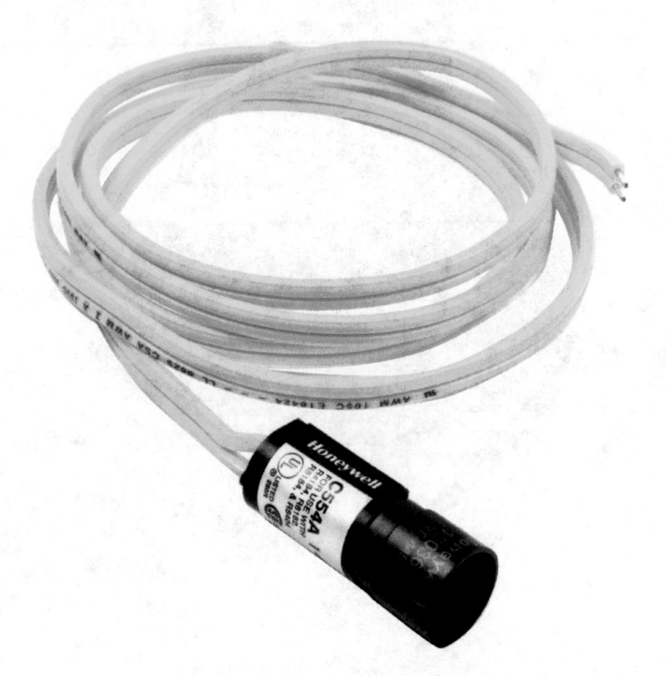

Figure 9-9. Typical cad cell

CHECKING CAD CELL PERFORMANCE

A bad cad cell may prevent the burner from lighting, or it may allow the burner to light normally but shut off after a few seconds because the resistance does not change, indicating that it has not seen the flame. Cad cell resistance varies with exposure to light, so an absolute check of cell performance requires the use of an ohmmeter. Figure 9-9 shows a typical cad cell.

Figure 9-10 illustrates the procedure for checking a light cell. Cad cell lead wires must be removed from the primary control. Then the burner is started and the F-F terminals are jumpered before the control can lock out on safety. If the terminals are jumpered before the start is attempted, the burner cannot be started. The jumper will bypass the safety switch, allowing the burner to run.

Connect an ohmmeter directly across the cad cell leads and measure the resistance. With the burner operating, cell resistance should be between 300 and 1,000 Ω if the burner is properly adjusted. The cell resistance may be as high as 1,000 to 1,600 Ω on a poorly adjusted burner. In any case, resistance must be less than 1,600 Ω for reliable performance. (Be aware that resistance can vary depending on the manufacturer. Always check specifications.)

After checking the light cell, check the dark cell resistance, as shown in Figure 9-11. This is to make sure that when no flame exists, the cell is not being affected by external light. Stop the burner and remove the F-F jumper. If the jumper is left in place, the burner cannot be started until it is removed. The resistance of the dark cell should be over 100,000 Ω.

There is little actual service that can be done on the cell itself. If it appears that there is an open circuit in the flame detector wiring, the cell should be checked for proper seating in its receptacle. Vibration could cause the cell to work loose. The cell surface may be cleaned with a soft cloth to remove soot or dust buildup that may be affecting its view of the flame. The plug-in portion of the cell itself may be replaced by simply removing the old cell and plugging the new one into the old receptacle. If the cell is replaced, the checkout procedure should be used to check its operation.

NOZZLE MAINTENANCE

A quality nozzle should last through a normal heating season if reasonably clean oil is supplied to it. There have been cases where a nozzle has worked several heating seasons if it is not overheated. However, experience has shown that the best results can be obtained (and obtained more economically) by replacing the nozzles annually. A nozzle should be reused only if a replacement nozzle is not available. Cleaning a nozzle properly is a painstaking, time-consuming job. At the lower flow rates, it is practically impossible to see

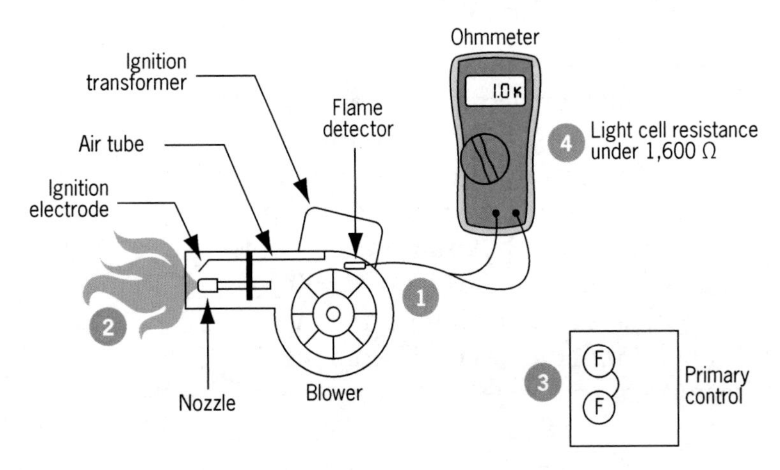

Figure 9-10. Checking cad cell (light cell)

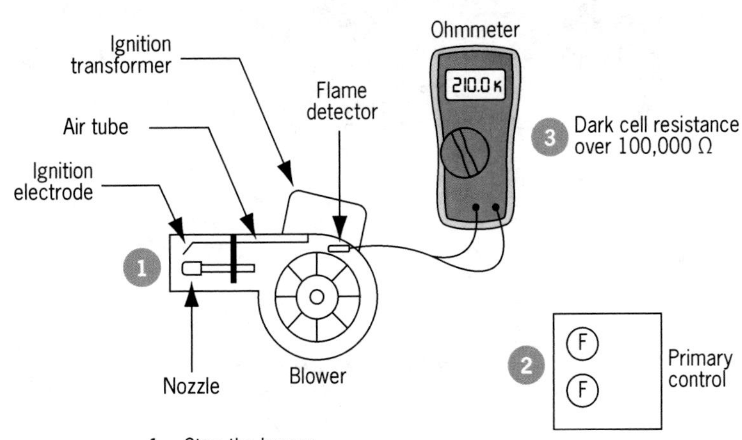

Figure 9-11. Checking cad cell (dark cell)

whether the distributor slots are thoroughly clean without a microscope. In the long run, time and money will be saved by a policy of annual nozzle replacement, rather than clean-up.

ELECTRODE TESTING AND SETTING

The manufacturer of the oil burner should be consulted for electrode gap settings and other settings for the oil burner. Many oil burner manufacturers have a gauge for adjusting the flame retention oil burner. Figure 9-12 shows an example of a multipurpose adjusting gauge. If the burner is an older burner and there is not information available from the manufacturer, the electrodes may be set according to the data shown in Figure 9-13.

Improperly located or gapped electrodes may cause a delayed ignition. The ignition electrodes are located in front of the nozzle, generally above the centerline. Electrode tips should never be permitted to touch or extend into the oil spray, because a carbon bridge will build up between them and ultimately cause ignition failure.

Ceramic insulators should always be treated gently. They should never be dropped or packed loosely in service kits. When servicing ceramic

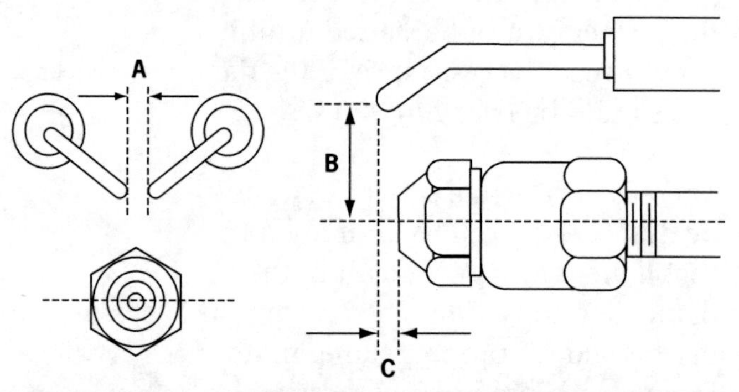

Nozzle	gal/hr	A, in.	B, in.	C, in.
45°	(0.75 to 4.0)	1/8 to 3/16	1/2 to 9/16	1/4
60°	(0.75 to 4.0)	1/8 to 3/16	9/16 to 5/8	1/4
70°	(0.75 to 4.0)	1/8 to 3/16	9/16 to 5/8	1/8
80°	(0.75 to 4.0)	1/8 to 3/16	9/16 to 5/8	1/8
90°	(0.75 to 4.0)	1/8 to 3/16	9/16 to 5/8	0

Note: Above 4.0 gal/hr, it may be advisable to increase dimension "C" by 1/8 in. to ensure smooth starting.

When using double adapters:

1. Twin ignition is safest and is recommended. Settings same as above.
2. With single ignition, use the same "A" and "B" dimensions as above. Add 1/4 in. to dimension "C." Locate the electrode gap on a line midway between the two nozzles.

Figure 9-13. Possible electrode settings

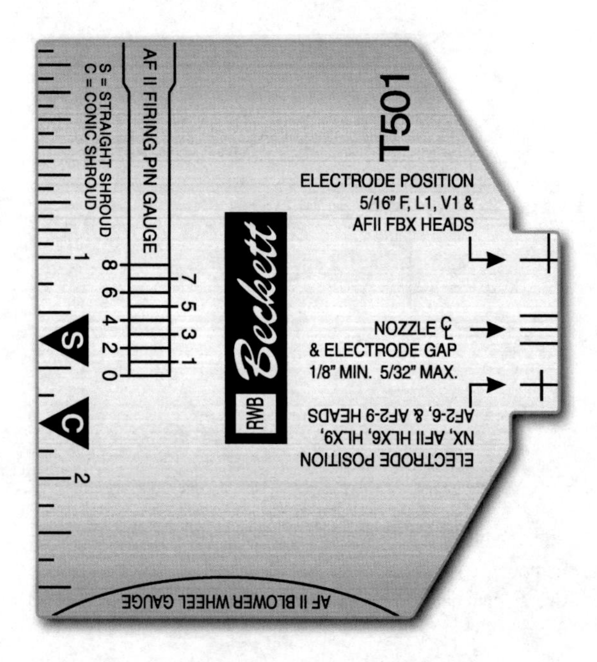

Figure 9-12. Beckett multipurpose gauge

insulators in the field, always wipe them clean with a cloth, or clean them with a solvent. If they show signs of aging or cracking, they should be replaced immediately. □

REVIEW QUESTIONS

1. What must a cad cell "see" in order for the ignition process to continue past the purge cycle?

   a. A dark combustion chamber
   b. A low resistance
   c. A proper oil pressure
   d. The presence of a spark

2. Carbon buildup on the walls of a combustion chamber may mean that the _________.

   a. burner is operating normally
   b. cad cell needs to be replaced
   c. firing rate needs to be adjusted
   d. oil pressure is too low

3. A properly adjusted flame retention oil burner should have a smoke reading of _________.

   a. 0
   b. 2
   c. 4
   d. 6

4. The *net stack temperature* can be determined by _________.

   a. adding the ambient temperature and the measured stack temperature
   b. averaging the ambient temperature and the measured stack temperature
   c. subtracting the ambient temperature from the measured stack temperature
   d. measuring the stack temperature before any draft adjustments have been made

5. Which of the following may be the cause of an abnormally high CO_2 reading?

   a. Air gate open too wide
   b. Not enough draft
   c. Nozzle too small
   d. Underfiring combustion chamber

6. An abnormally low stack temperature may be caused by a(n) _________.

   a. dirty heat exchanger
   b. excessive draft
   c. overfiring furnace
   d. underfiring furnace

7. A low CO_2 reading is an indication of _________.

   a. high oil pressure
   b. low oil pressure
   c. not enough excess air
   d. too much excess air

8. The purpose of a draft regulator is to _________.

   a. add dilution air
   b. add secondary air
   c. detect air leaks
   d. maintain a constant overfire draft

9. Which of the following conditions should be measured for proper combustion analysis?

   a. CO_2
   b. Smoke level
   c. Stack temperatures
   d. All of the above

10. Which of the following resistance readings indicates a dark cad cell?

    a. 100 Ω
    b. 500 Ω
    c. 5,000 Ω
    d. 200,000 Ω

11. How often should burner nozzles be replaced?

    a. Once a year
    b. Every 2 years
    c. Every 5 years
    d. Only when they can no longer be cleaned

12. What will happen if ignition electrodes are allowed to extend into the oil spray?

    a. Carbon will bridge the electrode gap
    b. The burner will smoke
    c. The oil will ignite too quickly
    d. The transformer will fail prematurely

ILLUSTRATION CREDITS
FIGURE 9-1: BACHARACH, INC.
FIGURE 9-2: BACHARACH, INC.
FIGURE 9-3: BACHARACH, INC.
FIGURE 9-5: BACHARACH, INC.
FIGURE 9-6: BACHARACH, INC.
FIGURE 9-7: BACHARACH, INC.
FIGURE 9-8: BACHARACH, INC.
FIGURE 9-9: HONEYWELL
FIGURE 9-12: BECKETT CORPORATION

APPENDIX A

Answers to Review Questions

INTRODUCTION

As you read and study the material in this book, you can test your understanding by answering the Review Questions at the end of each Chapter. Try to formulate your response *before* looking at the correct answer. If you make a mistake, go back and reread the relevant portion of the Chapter. □

Chapter 1: Introduction to Gas and Oil Heating

1. B	2. A	3. D	4. B
5. C	6. D	7. B	8. B
9. D	10. C	11. D	12. C
13. A	14. B	15. A	

Chapter 2: Combustion

1. D	2. B	3. B	4. D
5. D	6. A	7. A	8. A
9. A	10. D	11. C	12. C
13. B	14. D	15. A	

Chapter 3: Furnace Installation

1. B	2. A	3. D	4. B
5. B	6. C	7. B	8. A
9. C	10. A	11. C	12. A
13. B	14. B		

Chapter 4: Venting

1. A	2. D	3. D	4. A
5. C	6. B	7. B	8. A
9. D	10. D	11. C	12. D

Chapter 5: Air Flow

1. B	2. D	3. B	4. C
5. C	6. C	7. D	8. C
9. A	10. D	11. D	12. A
13. D	14. B	15. D	

Chapter 6: Troubleshooting

1. C	2. C	3. B	4. B
5. A	6. A	7. B	8. B
9. B	10. C	11. C	12. D
13. A	14. D	15. B	16. B

Chapter 7: Oil Burners

1. B	2. C	3. B	4. D
5. C	6. C	7. B	8. B
9. D	10. D	11. A	12. D
13. C	14. B		

Chapter 8: Oil Tanks

1. D	2. B	3. C	4. B
5. B	6. A	7. B	8. B
9. C	10. B		

Chapter 9: Planned Maintenance

1. A	2. C	3. A	4. C
5. B	6. D	7. D	8. D
9. D	10. D	11. A	12. A

ILLUSTRATION CREDITS
OVERLEAF: ISTOCKPHOTO

ACCA
Air Conditioning Contractors of America

PHCC
EDUCATIONAL FOUNDATION

RSES
The HVACR Training Authority

INDIVIDUAL MEMBERSHIP APPLICATION

RSES 1911 Rohlwing Road, Suite A Rolling Meadows, IL 60008-1397
Phone: 800-297-5660 or 847-297-6464 Fax: 847-297-5038 Web site: www.rses.org

FOR HEADQUARTERS USE ONLY

MC ______ Member No. ______________________ Chapter No. ________________ BC ________

PLEASE TYPE OR PRINT CLEARLY *the information you wish to be shown on all RSES records and correspondence.* ❑ Mr. ❑ Mrs. ❑ Ms.

First name ____________________ MI ___ Last name ____________________

Home address ______________________________ Apt.# ________

City ____________________ State/Province ______ Zip/Postal Code __________

Country ______________ Phone ______________ Cell ______________

Fax ______________ E-mail ______________________________

Month/Day/Year of birth ____________

RSES member within the last 3 years? ❑ Yes ❑ No Member No. ____________ CM ____ CMS ____

Send mailings to my: ❑ Home address (above) ❑ Business address (below)

Please check if you DO NOT wish to receive: Press releases, notices, announcements, and other information from RSES and the RSES Educational Foundation via: ❑ E-mail ❑ Fax

Business-related third-party offers via: ❑ Direct mail ❑ E-mail ❑ Fax

Having agreed to abide by the Society Bylaws, and those of any Chapter or subsidiary association to which I may belong, I hereby officially apply for membership in RSES.

Signature ______________________________ Date __________

EDUCATION/EMPLOYMENT INFORMATION

School Name *(most recent)* ______________________________ Years attended ______

City ____________________ State/Province ______ Area of study __________

Employer Name *(current or most recent)* ______________________________

Address ______________________________

City ____________________ State/Province ______ Zip/Postal Code __________

Country ______________ Phone ______________ Alt. Phone ______________

Fax ______________ E-mail ______________________________

PAYMENT METHOD

New Member Dues *(check one)* ❑ **One year: $128.00** ❑ **Two years: $230.40** *(save 10%)* ❑ **Three years: $326.40** *(save 15%)*

❑ Check enclosed *(make payable to RSES in U.S. dollars)*

Credit card ❑ VISA ❑ MASTERCARD ❑ AMERICAN EXPRESS ❑ DISCOVER

Card No. ______________________________ Expiration date __________

Authorized signature ______________________________ CCV __________

Note: A special *RSES Journal* individual subscription price ($18 per year) is included in membership dues. Members may not deduct the subscription price from dues. To determine if RSES membership dues are tax-deductible, consult your tax advisor.

ANSWERS REQUIRED

1. My primary HVACR role is: *(check one)*

- ❑ Contractor
- ❑ Service Technician/Installer
- ❑ Operations/Maintenance Manager/Engineer/Technician
- ❑ Engineer
- ❑ Sales
- ❑ Instructor
- ❑ Student
- ❑ Other __________ *(please specify)*

2. My firm's business is: *(check one)*

- ❑ Contractor: 1–3 technicians
- ❑ Contractor: 4–10 technicians
- ❑ Contractor: 11–19 technicians
- ❑ Contractor: 20+ technicians
- ❑ HVACR industry OEM
- ❑ Industrial (manufacturing or processing, not HVACR industry)
- ❑ Wholesaler/Distributor
- ❑ Commercial/Institutional/Government Agency/Association
- ❑ Other __________ *(please specify)*

3. I heard about RSES from: *(check one)*

- ❑ Seminar
- ❑ Chapter
- ❑ Friend
- ❑ Employer
- ❑ School/Instructor
- ❑ Member (current/former)
- ❑ Direct mail
- ❑ Internet
- ❑ *RSES Journal*
- ✔ Other EORD16 *(please specify)*

4. My e-mail preferences: *(check all that apply)*

- ❑ Conference
- ❑ Seminars
- ❑ Regional/Association/Chapter news
- ❑ General news
- ❑ Training and testing news
- ❑ Product news
- ❑ Chapter Officer news
- ❑ *RSES Journal* updates
- ❑ *RSES Journal* news and Web exclusives
- ❑ *RSES Journal* e-newsletters
- ❑ Industry news and events
- ❑ Membership benefits
- ❑ Business-related third-party offers